AF716302

Monetary Valuation of Environmental Impacts

Models and Data

Bengt Steen
Environmental System Analysis
Chalmers University of Technology
Göteborg, Sweden

CRC Press
Taylor & Francis Group
Boca Raton London New York

CRC Press is an imprint of the
Taylor & Francis Group, an **informa** business
A SCIENCE PUBLISHERS BOOK

CRC Press
Taylor & Francis Group
6000 Broken Sound Parkway NW, Suite 300
Boca Raton, FL 33487-2742

First issued in paperback 2021

© 2020 by Taylor & Francis Group, LLC
CRC Press is an imprint of Taylor & Francis Group, an Informa business

No claim to original U.S. Government works

Version Date: 20190826

ISBN-13: 978-1-138-36656-5 (hbk)
ISBN-13: 978-1-03-208624-8 (pbk)

This book contains information obtained from authentic and highly regarded sources. Reasonable efforts have been made to publish reliable data and information, but the author and publisher cannot assume responsibility for the validity of all materials or the consequences of their use. The authors and publishers have attempted to trace the copyright holders of all material reproduced in this publication and apologize to copyright holders if permission to publish in this form has not been obtained. If any copyright material has not been acknowledged please write and let us know so we may rectify in any future reprint.

Except as permitted under U.S. Copyright Law, no part of this book may be reprinted, reproduced, transmitted, or utilized in any form by any electronic, mechanical, or other means, now known or hereafter invented, including photocopying, microfilming, and recording, or in any information storage or retrieval system, without written permission from the publishers.

For permission to photocopy or use material electronically from this work, please access www.copyright.com (http://www.copyright.com/) or contact the Copyright Clearance Center, Inc. (CCC), 222 Rosewood Drive, Danvers, MA 01923, 978-750-8400. CCC is a not-for-profit organization that provides licenses and registration for a variety of users. For organizations that have been granted a photocopy license by the CCC, a separate system of payment has been arranged.

Trademark Notice: Product or corporate names may be trademarks or registered trademarks, and are used only for identification and explanation without intent to infringe.

Library of Congress Cataloging-in-Publication Data

Names: Steen, Bengt (Environmental scientist), author.
Title: Monetary valuation of environmental impacts : models and data / Bengt Steen.
Description: Boca Raton : CRC Press, [2020] | Includes bibliographical references and index. | Summary: "Linking environmental impacts from human activities, to the economy is generally advocated. But the non-availability of actual monetary values of impacts, hinders its use in practical terms. Another problem is the poor understanding of what is available, and the very different values it represents. The book describes how monetary values are determined, and provides a comprehensive set of long-term monetary values that may be used as default in environmental management with a focus on sustainability. Besides being a support to decisions, monetary valuation of environmental impacts offer an opportunity to learn about small and large environmental impacts"-- Provided by publisher.
Identifiers: LCCN 2019035752 | ISBN 9781138366565 (hardcover)
Subjects: LCSH: Environmental impact analysis--Economic aspects. | Environmental economics. | Environmental auditing.
Classification: LCC HD75.6 .S7425 2020 | DDC 333.71/4--dc23
LC record available at https://lccn.loc.gov/2019035752

Visit the Taylor & Francis Web site at
http://www.taylorandfrancis.com

and the CRC Press Web site at
http://www.crcpress.com

Preface

Models help us describe and understand reality. They also help in preparing for the future. Models are basically simplification of reality. Sophisticated models may be good at predicting the future but require extensive data collection and are difficult to explain. They become black boxes to many who are interested in their results. On the other hand, if the models are made too simple, they may not be entirely accurate. The models and data presented in this book are designed to be possible to grasp for most people with basic knowledge of mathematics, chemistry, physics and economics. They follow certain principles, and even if some more advanced reasoning is done for some issues, it does not hinder the use of the models.

Chapters 1–4 give the general background of models and examine a few models to see how they are constructed. This book can therefore be used as a reference tool when a value is of special interest. In Chapter 5 the monetary values of the elements of the natural capital (environmental goods) is determined. The models discussed in Chapters 6–14 link emissions and land use to environmental goods and can then be looked at one by one, when there is reason for it. There is as much as possible sufficient information in the description of each model. References to other parts of the text are kept to a minimum. Chapter 15–17 discuss the use and consequences of monetary values. Chapter 18 is a summary of values.

Acknowledgements

Funding from Vinnova and the Swedish Energy Agency is greatly appreciated and acknowledged.

The development of the models reported here have not been done in isolation. Informal exchange of ideas and data with colleagues is key to a creative process. A special thanks to Tomas Rydberg and OlleBoëthius for their patient listening and feedback.

I also want to thank all colleagues within the ISO working group developing the standard 14008. The opportunity to develop a framework for monetary valuation of environmental impacts in an international forum has been very valuable.

Contents

1

Introduction

It was a dark cold night of 18 January 1980. The bulk carrier *Star Clipper* was heading for the Uddevalla harbor on Sweden's west coast. The ship followed a route where it had to pass from under a bow-bridge connecting the mainland to the island of Tjörn. For some reason, which we shall never know, it strayed its course, and at 01.30 a.m. of the night of January 18 it hit the bow supporting the bridge. The bridge collapsed, fell on the ship, and crushed its communication radio. The ice prevented the people on the ship to reach land via the lifeboats. Eight people lost their lives before warnings could be set up. They drove off the broken bridge, into the ice cold water.

Twenty years earlier, on June 15, 1960, the bridge had been inaugurated after proper testing to see if it would be able to bear load of the traffic passing over it? Lots of trucks loaded with stones were parked on it, and measurements were taken. It passed all the tests and the bridge was considered safe and opened for traffic. It was a beautiful bridge that offered a stunning view of the sea and archipelago. I remember how on some weekends, we just drove over it just to enjoy the spectacular view. At that time it was considered a marvel of modern technology and progress. But it survived only 20 years and 7 months!

After a year, a new bridge was built. It was inaugurated in 1981. It had a different design. It had two towers placed on land on each side, and cables taking up the load from the traffic. The accident from 1980 could never happen again.

I often think of this incident, when struggling with the complexity of measuring sustainability. CES, the Conference of European Statisticians [1], initiated by UNECE, recommended between 24 and 90 indicators for measuring sustainable development. These were developed further into 17 UN sustainability goals with a lot of indicators to measure progress. In the bridge case, there was a unique and exact answer. The bridge sustained for 20 years and 7 months. And the dimension of sustainability was just one: time. Why can we not have such simple answers to all sustainability assessments, at least for environmental issues?

There are both differences and similarities between the bridge case and the environment. The main difference is that a bridge has only one function, while the environment has several functions. But for both, sustainability is about their carrying capacity. In a bridge's case the carrying capacity concerns vehicles. Its function is to allow vehicles to pass from one side to the other. In the case of the environment, the carrying capacity concerns goods and services supporting human welfare. The function of environment is to allow people to maintain welfare from one time to another. So, in both cases, the sustainability of the function should be possible to express in a temporal dimension, e.g., years.

Just as in the case of a bridge, the sustainability of the environment is determined by the most critical threat. In the bridge's case it was the ship that strayed from its course. For the environment it may be any of several threats such as extreme climate changes, depletion of the stratospheric ozone layer, or collapse of ecosystems. To determine the sustainability of the environmental function, one would therefore need to calculate the environment's capability of supplying sufficient goods and services for each threat for satisfying basic needs. Sustainability would then be determined by the threat with the highest probability of causing an insufficient service level for a basic need.

A more likely scenario than one, where people passively are watching their own destruction by a specific threat, is one where mitigation or adaption measures are taken. In the end, conventional economic resources will limit the ability for mitigation and adaption. The lesser the natural capital and available ecosystem services, the more conventional economic resources will be needed for mitigation and adaption. In fact both natural capital and conventional economic capital will be needed. In such a scenario, the environmental damage cost for those impacts that can be tolerated plus the cost to reach tolerable levels of those impacts that exceed tolerable levels would be a relevant indicator of the threat to environmental sustainability.

This is the chief reason why the common measure used in this book for impacts on the carrying capacity of the environment will be monetary, and that the monetary valuation of environmental impacts is (a) willingness to pay to avoid impacts that can be tolerated and (b) restoration costs for those that cannot be tolerated. What can be tolerated and not may change over time, but the more critical a resource is for satisfying basic needs and the more uncertain its function is, the less tolerable is a negative impact on the resource.

Such an economic measure is closely related to the concept of "natural capital". There are several international initiatives based on the concept of natural capital and its valuation. CES [1], has recommended measuring sustainable development by means of environmental, social, human and economic capital. G8+5 commissioned a study of "The Economics of Ecosystems and Biodiversity" (TEEB) to "make nature's values visible" (http://www.teebweb.org). The Natural Capital Coalition has developed a TEEB for business (https://naturalcapitalcoalition.org). ISO has developed a standard on monetary valuation of environmental impacts [2] and one standard on how to use environmental values in an organization's environmental management [3].

CES's recommendations builds on SEEA (system for environmental accounting), complemented with natural assets such as ecosystem and climate. They recommend including energy resources, mineral resources (excluding coal and peat), land and ecosystems, water, air quality and climate in assessing long-term sustainability.

The TEEB study identified three types of values (1) monetary, (2) quantitative, non-monetary and (3) qualitative, i.e., identifying something as valuable. Following the logic of sustainability being dependent on economic resources, (2) may be included in a monetary assessment if acceptable levels can be identified. Then calculating the costs to reach acceptable levels may be used as measures.

Monetary value of environmental goods and services are not carved in stone. They depend on the context. In case a unique and critical environmental good is threatened, its value increases drastically. So does the actions for sustainability, because in trade-offs between resource allocations, measures to preserve this particular environmental good will be prioritized. In that sense they are not different from ordinary market values that change with supply and demand.

The data and models shown in this book are the result of 30 years of development of a life cycle assessment (LCA) tool for designers, called the EPS system [4]. EPS stands for environmental priority strategies for product design. The tool was initially requested for by a designer working in the Volvo car company. He argued that there was a need for a "compass" showing which design option was preferable from an environmental point of view. It should be fast and approximately right. After some years of method development, we decided to use monetary valuations of the environmental impacts of the alternative designs. Then we could use similar calculations as is used for ordinary costing. Materials, processes and parts could have ordinary costs and environmental costs. If a material had a high environmental cost, it would directly trigger a thought of an alternative material or a design for recycling. This type of thinking could also be applied in other situations, e.g., purchasing, investments and marketing. So, besides being a relevant measure of threats to sustainability, monetary values are a practical tool in decision-making and most certainly, in future company profiling.

Monetary valuation can, however be made in several ways, and give different results depending on what is being valued, whose values that are expressed and on other contextual circumstances. The ISO standard 14008 offers a language and structure to specify what a monetary value of an environmental impact represents. The ISO 14008 standard will be used in describing monetary values of environmental impacts in this book.

References

1. *Conference of European Statisticians Recommendations on Measuring Sustainable Development*. 2014, UNECE: New York and Geneva. p. 226.
2. ISO, *Monetary valuation of environmental impacts and related aspects*. ISO 14008:2019, ISO, Geneva.
3. ISO, *Environmental management—Determining environmental costs and & benefits* ISO 14007:2019, ISO, Geneva.
4. Steen, B., Calculation of Monetary Values of Environmental Impacts from Emissions and Resource Use. The Case of Using the EPS 2015d Impact Assessment Method. *Journal of Sustainable Development*, 2016. **9**(6): p. 18.

2

Goal and Scope

There is no single, universally applicable monetary value for an environmental impact. For a commercial company, their customers' valuation of their product's environmental performance, expressed by the information customers get, may be most relevant. For instance, such information may be emissions of climate gases, expressed as CO_2-equivalents. For a government trying to reach certain targets, environmental impact abatement costs may be of highest relevance. For long-term planning, valuations that are constant over time, or at least foreseeable, such as for some marketed goods, may be most relevant. The ISO 14008 standard on monetary valuation of environmental impacts is therefore very strict on documentation and reporting of the goal and scope of any value assessment. The goal and scope of the valuations reported in this book is described below according to the standard requirement.

2.1 Goal

The goal of this book is to determine and document monetary values of environmental impacts from emissions and use of natural resources. These values are intended to be used for determining environmental impact costs for products and services. The environmental impact cost for products and services are in turn intended to be used to indicate which of the two alternatives is most sustainable in everyday choices between product or service alternatives. Similarly, as ordinary costs are available, and used for choosing between alternatives, monetary values of environmental cost can be made available and used for choosing between the alternatives. It should be noted however that much of these costs are costs to future generations and should not be integrated in the current economy without reflection.

The intended audience of this book are product developers, policy makers, purchasers and consumers. Their knowledge of environmental issues may vary from that of an ordinary consumer to expert level.

The reason for focusing on everyday choices is that our life conditions and their sustainability to a very high degree are shaped by human decisions. There is of course a lot of impact on our life conditions from processes that we cannot control, but that only puts an extra pressure on making informed choices where we can make a difference.

Monetary values of emissions and resources can be determined in several ways and the values vary depending on what is included, whose values that are expressed and what is assumed about the future. The monetary valuation documented in this book will follow the ISO 14008 Standard on monetary valuation of environmental impacts and related environmental aspects to ensure that there is transparency on how they are calculated. The models and monetary values determined can be regarded as a kind of default values for global, inter-generational impacts. If conditions in a specific case is different from the default scenario, it is easier to change these input data and find a new more relevant monetary value, than to start all over from the beginning.

Hopefully this book will not only offer some useful data to the readers, but also trigger reflection and perhaps objections about the monetary values assessed. It is only through being concrete that one may get concrete feedback. In a field with so many qualitative statements, there is a need for numbers.

The ISO 14008 Standard is a framework standard and includes definitions of terms, guiding principles, planning, doing and reporting. In the planning the scope is set with some guiding principles in mind.

2.2 Scope

The guiding principles of ISO 14008 are accuracy, completeness, consistency, credibility, relevance and transparency.

2.2.1 Accuracy

Scientific literature and reports from renowned organizations are the preferred sources of information. External data are referenced as much as possible and relevant data about data documented. Despite this, there is a lot of uncertainty. In fact, uncertainty is a part of reality and always present in some way or another. Some of the uncertainty is due to real variations, such as impact of emissions from different sources at different times. Some are due to lack of knowledge, such as uncertainty caused by an unknown future.

Data therefore always have two figures: a best estimate and a measure of uncertainty. If nothing else is documented, a log-normal distribution is assumed. The best estimate is the average logarithmic value, which in a log-normal distribution is equal to the median of the original value estimates. In case of a log-normal distribution, uncertainty is described as a factor by which the median value may be multiplied or divided to find the values representing one standard deviation higher or lower values.

2.2.2 Completeness

The completeness principle in ISO Standard requires including significant information. What is significant is revealed only when the information is used in a choice situation. However, information on known environmental threats to human welfare is included as default. The completeness principle concerns several issues in the determination of monetary values of emissions and use of resources: which emissions and resources to include, and which environmental goods and pathways (mechanisms) to include. To find what is significant at a general level, the Brundtland definition and its focus on basic needs is a good starting point. Severity and awareness of environmental threats to satisfiers of basic needs indicate (1) which environmental goods and mechanisms and (2) which emissions and resource use to include (Figure 2.1).

One particular problem in valuing environmental impacts and resource use, has to do with uncertain knowledge. When should the data be regarded as reliable? If requiring high accuracy and scientific rigidness,

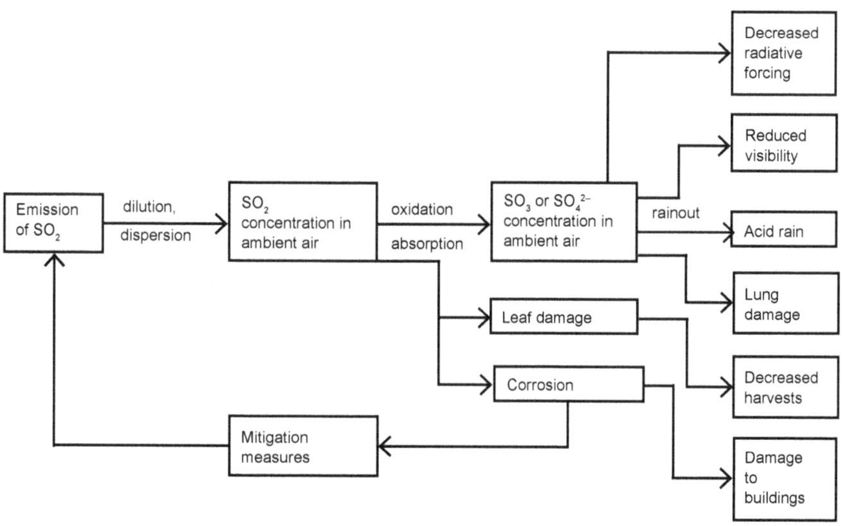

Figure 2.1. Part of cause-effect mechanism for SO_2.

several impacts should not be included in the valuation and let their value in a practical choice be zero. So, if focus is scientific rigidness in the valuation, early uncertain knowledge should not be included. But if focus is on sustainable products even uncertain knowledge will be reasonable to include. As our goal is to support choices for sustainable products and services, any data or causality that has been published in scientific literature will be included, although full consensus may not have been achieved.

In trying to be complete, there is a risk of double counting. For instance, impacts on the production capacity of irrigation water may be accounted for once through value of decreasing irrigation water and once by decreased production capacity for crops. The approach here is not to value supporting environmental goods, like irrigation water, if the links to basic need satisfiers are known. If they are not known, like for instance biodiversity, they are valued separately.

2.2.3 Consistency

Sometimes, in methodological choices, there may be a conflict between consistency and accuracy. The standard methodological approach may not be the best choice. However, in order to be able to explain the methodology to a large audience of users, consistency is favored, even if there are genuine reasons for an alternative approach in a specific situation. It is believed that it is better to address a possible bias from such a choice in a sensitivity analysis, than being inconsistent with which methodology to use.

The monetary values determined are all for global averages 2018, WTP by people in OECD, and with 0% discounting rate.

2.2.4 Credibility

Using scientific material, applying international standards and referencing all data and models are methods used to enhance credibility.

2.2.5 Relevance

By focusing on real choices, including uncertainty and sensitivity analysis, the data and models most important for achieving significant differences between alternatives can be identified and gradually improved. Normally, there are only a handful of data and models that account for 80% of the monetary value of the environmental impacts from a product or service. It is important that these data and models be as much good as possible. Other models could also be useful, even if their accuracy is low. It would then be sufficient to know the magnitude of their impact values.

2.2.6 Transparency

Transparency is created by making all data and calculations traceable, and also through simplification and consistency. Simplification should preferably be made by showing relevant data to each user and keeping the background data traceable and not omitting the background data.

2.3 Planning

In ISO 14008, the planning phase is when the goal and scope of the study are precisely defined and documented. This includes the currency and its base year, whose values are being expressed, what is being valued, discounting and how uncertainty will be addressed.

The currency of the monetary value should be given in US Dollars and the base year must be 2018. The values for changes in environmental state will typically correspond to one year of change, e.g., one year of a reduced health status for a person. Sometimes the monetary value will be for a one-off change, like one less kilo of crop produced. The reference unit of monetary value will be per individual.

In a choice situation, it is interesting to know what the person considering alternatives will be willing to pay (or accept compensation) for the change caused in the environment if it impacts him/her.

For each aspect (emission or resource use), the monetary value will be aggregated over time, space, people, and environmental impacts. Changes in the environment will be valued for as long as they persist, globally, for all people that are concerned and for all relevant impacts.

Currency and base year adjustments can be made as per requirements in ISO 14008. Adjustments for income or equity weighing will not be made as all monetary valuations are either world market prices or related to OECD average productivity.

Discounting of future impact values will be made with 0%. The reason is that the dominating contributions to monetary environmental impact values comes from health effects and resource depletion. WTP for health is assumed to follow productivity and income. Following the "polluter pays the principle", depletion of natural resources is accounted for when they occur, which normally is the same time as they are extracted from the environment.

Uncertainty is estimated in two steps. First, a distribution is chosen. Here log-normal distributions are used by default. Second, a standard deviation is estimated from real variations and from model uncertainty. Sensitivity analyses are not made on monetary impact values, but data are compiled in such a way as to allow sensitivity analysis when applying monetary impact values in choices between product or service options.

The monetary values determined will be the average or median value for the space and time where it occurs. But as the models are linear, the same values apply for marginal changes also.

2.3.1 Specification of the environmental impacts, emissions and resource uses to be valued

In literature, monetary values estimated for human activities, e.g., driving a car for 1 km, or producing 1 kWh of electricity, often differ significantly. Even for single emissions, for instance, CO_2, monetary values may vary several orders of magnitude [1]. One of the main reasons is that different emissions, resources and impacts have been included in different studies. The choice of impacts, emissions and resources is critical for the value obtained.

But how can we find out which impacts, emissions and resource uses are relevant to the value?

The guiding principle of completeness indicates that everything should be accounted for. Reviewing the scientific literature will give us a long list, including impacts close to the source as well as impacts induced by environmental mechanisms in several steps. Figure 2.1 shows a part of such cause-effect mechanism for SO_2. In LCA practice, one often speaks of mid-point and end-point impacts [2]. Mid-point impacts could be "acidification potential", while end-point impacts could be lung damage, damage to buildings and decreased harvests. In real life, there are no end-points. Every change causes other changes, but when modelling, there must be end-points in the models.

Impacts are a bit like concentric rings on water when we throw a pebble. The further away you are from the source the more kinds of impacts may occur. Sometimes environmental impacts generate new human activities, which cause more emissions. The cause-effect chains are not always restricted to environment, although it is rare that one can model cause-effects that have pathways returning to technical-economic processes. An exception in this book is corrosion of steel.

The long list of impacts may be shortened by only including impacts relevant to sustainability. Environmental threats to satisfiers of basic needs define which of the environmental impacts it is relevant to include, and which emissions and resource uses are responsible. In this way principles of completeness can be met.

There are many things in the environment that are temporal in a long-term perspective, such as scenic beauty and spiritual values, and have a vague impact on sustainability. In this book, the Brundtland definition of sustainable development has been used. This means that it is human welfare that is to be developed and sustained. In the well-known report

"Our Common Future" the focus is on basic needs, although other needs are also mentioned [3]. A sentence in this document, "Sustainable development requires meeting the basic needs of all and extending to all the opportunity to satisfy their aspirations for a better life" indicates this.

The principle of equal rights to all generations has been applied in this book. In a not too pessimistic scenario, future generations are in clear majority. This means that environmental goods specific to the present culture, such as sea-shells or diamonds for jewelry, are of minor relevance to value. The Brundtland definition focuses on human welfare, alleviation of poverty and basic needs. Basic needs are underpinning welfare independent of culture and will remain as long as human beings exist. It is therefore logical to **focus on impacts, emissions and use of resources that are known to threaten satisfiers to basic needs**.

The basic needs concept was originally made public by the International Labor Organization (ILO) [4]. Since then, several lists of basic needs have come up [5–9]. They vary in what they are called and in what are called needs and satisfiers. In Table 2.1 one set of needs is listed, which are believed to be relevant for life-cycle sustainability assessment (LCSA) of products. There are also several psychological needs that could be of interest, but they have been omitted here, as it is not likely to find good models to show how they influence sustainability and how relevant they are to future generations. The list is chosen to give as little redundancy as possible. The exact names of the basic needs are probably not so critical in

Table 2.1. Basic needs.

Need	Function
Nourishment	Source of energy for life functions and building materials for the body
Water	Solvent, a component in metabolisms, heat control
Oxygen	Used for energy release when combining with food
Excretion	Exhalation, urine, feces, sweat
Proper temperature	Certain temperature range to maintain living processes
Body Integrity	Protection from physical and psychological intrusion and from infections and toxic substances
Reproduction, sex	Enables future generations
Sleep, rest	Recovering
Movement	For health and for satisfying other needs
Identity	Ensures social belonging and role
Planning	For the management of present and future needs satisfaction
Love, attachment	For personal development and cooperation

identifying impacts, emissions and resources to value, but it is important that no relevant basic need is left out.

Satisfiers are resources that can be used to meet needs. Satisfiers are closely linked to basic needs but can be subjects for some innovation. New things to eat or new ways of finding an identity can develop. Creating identity through consumption is often questioned. For all satisfiers there is a supply chain, where several factors may determine the amount of satisfier that is available.

The process used here to identify satisfiers starts with reviewing literature on the nature of needs and how needs are satisfied today. Then, a relevant name is chosen including one or more types of satisfiers. I have tried to keep the list of satisfiers as short as possible in order to allow an easy overview.

The satisfiers identified for different basic needs are shown in Figure 2.2.

For nourishment the satisfier is normally *food* of acceptable quality. Although nourishment could be supplied directly into the blood, like in hospitals, food is the dominant satisfier. Food adds energy and are the

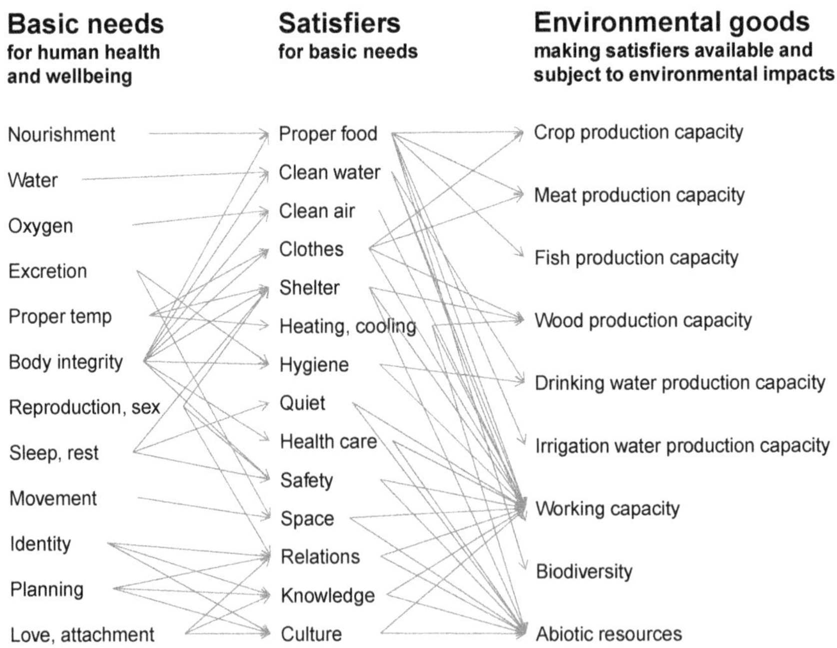

Figure 2.2. Identified satisfiers for basic needs and their corresponding environmental goods, subject to environmental impacts.

building blocks of the human body. Energy may be measured in terms of calories, while measures of building blocks are more diverse, like proteins, vitamins, minerals, etc.

For water the satisfier is *clean water*. Drinking water must be of a certain quality. Requirement on water quality is given by WHO. The satisfier for oxygen is *clean air*. Like for water, WHO issues quality guidelines for air. The satisfier for excretion is primarily available *space*. Hygiene installations like toilets and showers are considered basic in most cultures. Proper temperature may be satisfied through *shelter*, *heating/cooling* and through *clothing*.

There are several satisfiers for body integrity: *proper food, water, air, clothing, shelter, hygiene, healthcare* and *safety* from violence. Reproduction is a very basic need for any species to sustain. *Shelter, safety*, and *relations* are some of the satisfiers for reproduction. *Shelter, safety* and *quiet* environments are also satisfiers for sleep and rest. Movement requires *space*.

Identity is developed through *relations, knowledge* and *culture*. Relations, knowledge and culture are also satisfiers for planning, love and attachment.

Having identified the satisfiers shown in Figure 2.2, the next step is to identify which environmental goods that are needed for making satisfiers available and which are subject to environmental impacts.

Proper food requires several ecosystem services and some human work to get food on the table. Ecosystem services may be described in terms of supplying, supporting, regulating and cultural services. As pointed out by Sen [10], it is **capabilities** that is of value for sustainability rather than production per se. Systems are seldom threatened by average conditions, but by extreme events. Then, the capability of extra production is a good measure of resilience, not the production that are used. Ideally, we would like to be able to value surplus of environmental goods, i.e., capability minus use. But as monetary valuation anyway values changes in state, the same values will apply for resource surplus and capability. Changes in capabilities will be the same as changes in surplus.

Getting food on the table requires transportation, storage, food processing and other technologies, which is dependent on working capacity, access to water and abiotic resources.

Ecosystem services and human working capacity for making food available are threatened in many ways by environmental impacts from human activities. The most common concerns are shown in Table 2.2.

Clean water requires production capacity of drinking water and/or irrigation water, abiotic resources, and some human working capacity.

Clean air requires abatement technology, which in turn requires abiotic resources and working capacity.

Table 2.2. Common concerns of environmental impacts on environmental goods.

Environmental goods making satisfiers available and subject to environmental impacts	Environmental impacts
Crop production capacity	Soil degradation, air pollution, climate change, land use
Fish production capacity	Acidification, eutrophication, climate change, toxicity
Meat production capacity	Soil degradation, climate change, land use, toxicity
Wood production capacity	Climate change, air pollution, land use
Drinking water production capacity	Climate change, land use, water pollution
Irrigation water production capacity	Climate change, water pollution, land use
Working capacity	Health effects from climate change, air pollution, toxicity
Biodiversity	Land use, toxicity
Abiotic resources	Mining

To produce *clothes*, fibers and leather are needed, e.g., from sheep and cows, or from crops or wood.

Shelter requires building material, such as wood and abiotic resources and some working capacity.

To get a proper temperature *heating* or *cooling* may be needed, which may use fuels such as wood, biomass and abiotic resources besides some working capacity.

To promote good *hygiene*, water of good quality, i.e., drinking water is needed as well as some arrangements requiring working capacity.

A *quiet* surrounding may require some arrangements, which in turn requires working capacity and abiotic resources. The same is relevant for *health care, safety, space, relations, knowledge* and *culture*.

When identifying environmental goods to evaluate, the concept of "safeguard subjects" [11] or "areas of protection" [2] are often used for guidance. The ISO 14040 LCA standard has a guiding principle of comprehensiveness, saying that "LCA considers all attributes or aspects of the natural environment, human health and resources". The ILCD handbook on LCA [2] requires impacts on human health, natural environment and natural resources to be considered. The list of impacts valued in this book is limited to impacts on environmental goods making satisfiers for basic needs available. This means that impacts on human health is limited to DALY, disability adjusted life years. It is the ability to "work" that is valued, not impacts on human health in a broad sense, as human health is defined by WHO. The WHO health concept is here seen as the overall goal of sustainable development. It also means that impacts on natural environment is limited to those on the production capacity of crop, meat, fish, wood, drinking water and biodiversity. Impacts on resources are limited to those on abiotic resources.

Having defined what kind of impacts to value, we now need to decide what kind of emissions and resource use to assess.

All monetary values will be global averages or medians and refer to one extra unit of emission or resource extraction. Emissions known to contribute to climate change, acidification, eutrophication, stratospheric ozone depletion, ground level oxidant formation, toxic effects, secondary particle formation, fertilization or otherwise impacting on the environmental goods identified above will be assessed. But only emissions known to cause actual environmental impacts on the goods listed above are assessed. This means that emissions of many substances known to be potentially toxic has been left out. If they exist as trace amounts in exhaust fumes of vehicles, they cause no or insignificant impacts on the environmental goods identified here as an average. Also, emissions that only cause impacts on other environmental goods than the one identified here has been left out, e.g., short-term nuisance to humans or increased metal concentrations in marine life.

2.3.2 People whose preferences and perspectives are considered

Here, as default, the preferences and perspectives of an average OECD citizen is applied. There are several reasons for this:

- OECD countries are main decision-makers for changes to sustainable development
- The change to sustainability is largely an ethical issue, where it is relevant to present to the decision-maker what he or she would think about the impacts caused if it impacts him or her
- Monetary values for many environmental impacts are available in literatures for OECD conditions

In classical economics, future changes are normally translated to net present value. Valuations are associated by living people, who can express their values. In a sustainability context it is relevant to consider the economy of future generations. Of course, we cannot measure their values, but we can use our own values for future scenarios. What would we think of the impacts occurring in the future if they affected us? So, even if the preferences and perspectives used are present OECD citizens, they are used to represent all persons affected by an impact from emissions and resources use, even if the impacts last several generations.

2.3.3 Elements of the total economic value considered

The ISO 14008 standard requires specification of which elements of the total economic values that is intended to be assessed. The total economic value consists of use values and non-use values. The use values may be subdivided into direct use, indirect use and optional values. The non-use values may be subdivided into existence values, bequest values, and altruistic values. The assessments made in this book is mainly of use values, even if they are made for future generations and the global population. In that sense, there is also bequest and altruistic values.

2.3.4 Monetary valuation methods

Goods like crops, meat, fish, wood, water and labor are traded in the market and their monetary values can be found from various statistics. Even if the environmental goods we value are defined as capability for production, we value changes in capability, which are units of crop, meat, fish, wood, water and labor. Producer prices in the market are used as proxies for environmental damage costs per unit good. Producer prices are more suited for estimating value losses per unit good from environmental damage than consumer prices. Costs for farming, fishing, etc., are about the same with and without environmental damage, which results in less value created per unit good, while transports, processing and marketing costs do not change per unit good.

A decrease of a stock of abiotic resources are valued through the cost for its restoration with a sustainable alternative. There are three reasons for using this valuation method:

1. In the long run, this is a very likely scenario. Even if not restored now, future generations will have to find alternatives that are sustainable. As future generations—in a sustainability perspective—are in clear majority, "people whose perspectives and preferences are considered" will value concentrates like our ores at least as high as their cost for a sustainable alternative.
2. It is practical. If the value loss is accounted for when a resource is decreased, then it is independent on uncertain future scenarios.
3. It is advocated for in literature and was proposed already in the 1920s [12].

Biodiversity is almost impossible to value. Not only is it a complex good, which is difficult to measure, its quantitative relations to other environmental goods is largely unknown. The only monetary measure that can be estimated is the cost of prevention of declining biodiversity. This valuation method is used here.

References

1. Isacs, L. et al., Choosing a monetary value of greenhouse gases in assessment tools: A comprehensive review. *Journal of Cleaner Production*, 2016. **127**: 37–48.
2. *ILCD Handbook: General guide for Life Cycle Assessment - Detailed guidance*. 2010, European Commission - Joint Research Centre - Institute for Environment and Sustainability: Luxembourg. p. 417.
3. *Our Common Future*, in *UN Documents: Gathering a Body of Global Agreements*. 1987, UN.
4. Jolly, R. *Futures 8:* in *The world employment conference: tripartite World Conference on Employment, income distribution and social progress and the International Division of Labour* 1976. Geneva: International Labour Office.
5. Lütkenhorst, W., Operationalizing basic needs: a few fundamental reflections. *Intereconomics*, 1982: pp. 244–250.
6. Max-Neef, M., *Desarrollo a Escala Humana*, in *Editorial Nordan-Comunidad*. 1993: Montevideo.
7. Galtung, J., *Peace by peaceful means*. 1996, London: SAGE Publications.
8. Litwack, L., Basic needs—a retrospective. *Int J Real Ther*, 2007. **26**: 28–30.
9. UNDP, *Human development report 1997*. 1997, UNDP: New York Oxford.
10. Robeyns, I., The Capability Approach: a theoretical survey. *Journal of Human Development*, 2005. 6(1): 93–114.
11. Beltrani, G., Safeguard subjects. *Int J Life Cycle Assess*, 1997. **2**: 7.
12. Daly, H.E., *The Steady State Economy*, in *Valuing the Earth*, H.E.D.a.K.N. Townsend, Editor. 1993, Massachusets Institute of Technology: Cambridge, Massachusets. p. 344.

3

History and Outlook

A healthy environment is no doubt a fundamental prerequisite for sustainable development. But there are also others. The three pillars (social, economic and environmental), triple bottom line (people, planet, profit), and the four capitals (human, social, natural and economic) are conceptual models for what is needed for a sustainable development.

There are several reasons to quantify environmental or sustainability performance. Historically, environmental issues came on the agenda because of local violations of air and water quality. The London smog episode of 1952 and the Minamata catastrophe of 1956 are examples of this. It was then natural to look for the guilty polluter. Most environmental assessments from that time aimed for estimating which part of the problem that could be attributed to some defined activities. Greenwashing became a common term. At the end of the 20th century and beginning of the 21st, it was realized that industry was not only a cause of the problems but also the potential solution. There was a business case for sustainable development [1]. Then the consequences of an activity or change became more interesting than before.

Before 1985 the focus on environmental issues was emissions from stacks and pipes, and depletion of natural resources. In the late 1980s focus shifted to products. I remember especially one picture that showed a factory making organic solvent, and a truck leaving the gates with the product, while a lot of people were busy measuring the emissions from the stack.

When analyzing environmental impacts from products, there is a supply change that must be included to have the full picture of the production of a product. And then the step is not long to include the us and end-of-life phases. Life cycle assessments (LCA) were born. The development of LCA was stimulated by the introduction of the

PC and spreadsheets. Before 1990 it was practically impossible to make anything else than very simple LCAs. My first product oriented environmental assessment was in 1969. I was measuring the HCl formed when incinerating a PVC bottle. The results had to be calculated with a mechanical machine. Each multiplication took a minute or two. And there were many. Besides changing the possibilities for handling a lot of data, the internet has drastically changed the possibilities of exchanging data and other information. In the 1990s, I had to walk to the library and borrow books and papers to find what I was looking for. Now everything is more or less a available on our fingertips, thanks to the internet. Not only is it easy to download, synthesized information is abundant. Environmental research has grown immensely and set its mark in the scientific literature.

The mainstream development strategy for life cycle impact assessment, is positivistic in character. It asks the question: which are the environmental impacts of emissions and use of natural resources for a defined human activity? It follows these flows (called elementary flows) from the activity through the environment and ultimately to some point where the impact is characterized. When doing that, mainstream LCA tries to stick to natural science as much as possible. The current debate is very much on making robust theories linking elementary flows to impacts via characterization factors. ISO's LCA standards (ISO 14040 and 14044), and several guidebooks on LCA use this strategy [2]. The interpretation phase, when you interpret the meaning of the different impact category indicators are optional steps, considered subjective, and largely left to the user of the results.

Our development approach is hermeneutic in character. It asks the question: which of two alternatives is the most sustainable? To answer that question, all impacts must be described in one dimension and aggregated to one number. Any flow or impact can then be described in terms of how much it contributes to that number or the difference between the alternative choices. This allows a hermeneutic development in that it is possible to identify those data and models that need to be improved to improve the overall robustness in choosing between alternatives. If mainstream LCA is natural science, our approach is engineering science. It is the choice that shall be robust. Theories on impact characterization is a means but not a goal. In practice it means that any correlation between an elementary flow and an environmental impact is valuable knowledge. The requirement for precision in data and characterization models depends on how much they contribute to the robustness of the choice.

Since the early 1990s, it became clear to me that a monetary valuation of environmental impacts was a good way to obtain one number. The total environmental impact value in monetary terms was something that product developers would understand. Determining monetary values of environmental impacts from emissions and resource extractions are however not easy. In the beginning, I, and presumably other researchers in the field, was fully content if I just found a reasonable number for the value. Later, when different monetary values were compared, it was obvious that they differed considerably and that it not always was clear why. A harmonization was needed to create a common framework for monetary valuation. Preparations for an international standard on monetary valuations started in 2014 and the standard was published in March 2019. The standard, ISO 14008:2019 Monetary Valuation Of Environmental Impacts And Related Environmental Aspects, offers a common framework that ensures transparency and consistency in monetary valuation.

The standard is a key to common databases and softwares, which are needed for a broad use of monetary values of environmental impacts. The LCA standards ISO 14040 and 14044, did just that. They were framework standards that stimulated cooperation between experts. Today tens of thousands of people use LCA softwares and databases.

ISO 14008 was the first ISO standard in environmental economy. Now several new standards in this area is being developed and an entirely new technical committee on "sustainable finance" has been launched. Once the numbers are determined, the accounting can begin. Several companies, e.g., Puma and AkzoNobel, use monetary values of environmental impacts for sustainability reporting.

References

1. Heemskerk, B. et al., *Sustainable Development Reporting: Striking the balance*. 2002, World Business Council for Sustainable Development: Conches-Geneva.
2. *ILCD Handbook: General guide for Life Cycle Assessment - Detailed guidance*. 2010, European Commission - Joint Research Centre - Institute for Environment and Sustainability: Luxembourg. p. 417.

4

Terms and Definitions, Abbreviations

As much as possible, terms used in this book will correspond to their normal meaning, as defined in Webster's dictionary. But for some terms, more precise meaning relevant for monetary valuation of environmental impacts is needed. In those cases, the definitions of the ISO standard 14008 will be used. Some of the central ones are described below.

4.1 Terms and Definitions

4.1.1 Elementary flow

Material or energy entering the system being studied that has been drawn from the environment without previous human transformation, or material or energy leaving the system being studied that is released into the environment without subsequent human transformation (from ISO 14044).

4.1.2 Good

Natural resource, ecosystem service, product or service, marketed or not, that satisfies human wants or needs.

This term is very central as it represents the things we value. It could be an object, it could be a service, or it could be a capability. Environmental economics usually value marginal changes in flows of utilities. In a sustainability context it would rather be the capability to deliver goods that is of interest and not the deliverance as such. The most severe threats to ecosystems and human welfare do not come from an average situation, they come from extreme situations.

In cases where it is unclear what kinds of goods that is meant, it may be clarified as environmental good or consumption good.

Outside this book, differences are made between goods and services. However, the distinction is not always clear and instead of always repeating "goods and services" one may as well just use the term "good".

4.1.3 Natural resource

Part of nature which provides benefits to humans or underpins human well-being.

An apple is not a resource if you do not know that you can eat it.

4.1.4 Ecosystem service

Benefit people obtain from *ecosystems.*

4.1.5 Environmental impact

Change to the *environment,* whether adverse or beneficial, wholly or partially resulting from an organization's *environmental aspects.*

This definition is given in the ISO 14001 standard. In that standard the focus is on an organizations environmental impacts. In this book, there is a slightly different meaning given to that term. The change to the environment does not have to be caused by an organization, it may be caused by any emission or use of a natural resource.

4.1.6 Environmental impact factor

Quantity of environmental impact per quantity of emission or resource flow.

4.2 Abbreviations

BEV	Battery Electric Vehicles
BOD	Biological oxygen demand
CBA	Cost-benefit analysis
CES	Conference of European Statisticians
COD	Chemical Oxygen Demand
EPS	Environmental Priority Strategies for product design

IEA	International Energy Agency
ISO	International Standardization Organization
LCA	Life cycle analysis
LCIA	Life cycle impact assessment
LCSA	Life Cycle Sustainability Assessment
SEEA	System for Environmental-Economic Accounting
TEEB	The Economics of Ecosystems and Biodiversity
UNECE	United Nations Economic Commission for Europe
WTA	Willingness to accept (compensation)
WTP	Willingness To Pay
YLD	Years Lived with Disability
YLL	Years of Lost Life expectancy

4.3 Mathematical expression

Calculations explained in the text is written in Microsoft Excel format. Multiplication are represented by *, instead of the more common x or ·. Exponential numbers are written like 3E+05, instead of the more common $3 \cdot 10^5$.

5

Monetary Values of Environmental Goods

As described in the chapter on Goal and Scope, market values are used to estimate the monetary values of changes in the ecosystem production capacity and human working capacity. Please note that renewable use of environmental goods, like drinking water is not accounted for, only the capability to provide the good. The capability of providing drinking water by, e.g., a well may be decreased either by decreased flow or by decreased quality, making the water unsuitable for drinking. In case there is extraction from a non-renewable stock, every extraction is a decrease of its capability to provide its environmental good. In a sustainable society, the non-renewable stock must be replaced by a sustainable alternative. That replacement cost is a measure of the long-term value loss of the non-renewable stock. In a way, the replacement cost is a lower estimate of the market value for future generations. Future generations, not having easy access to the good, will be willing to pay at least as much as it costs to produce the sustainable alternative. In the long run, future generations are in clear majority. This means that their values are representative for those concerned (= all generations together) by the marginal changes in non-renewable environmental goods, not our present values.

Market values are used both for impacts on traded and non-traded goods. Health impacts on the non-working population is valued equally as impacts on the working population. Labor costs are a measure of the value of working capability. The working capability exists during all active years and during all waken hours, and are used, or optional for use, to sustain welfare. Market prices are used as proxies for value, not as the value itself. Market prices are normally a lower estimate of monetary values. Many buyers are willing to pay more than the prices, but there are seldom any data on this, the so called "surplus value".

Biodiversity is particularly tricky to value. Biodiversity has several values. It is a genetic bank, it strengthens ecosystem resilience and it supports ecosystem services. Present knowledge is not sufficient to allow quantitative modelling of the links between biodiversity characteristics and satisfiers to human needs. The role of biodiversity for ecosystem services is, at most, known for single issues, such as the threat to pollinators. Therefore, biodiversity is valued by the costs of conservation measures to preserve it on the level which today is implied or deemed necessary.

Valuing biodiversity in total, does not help when evaluating single human activities, such as land use. Then, the IUCNs databases on red-lists are used to allocate the total value to specific human activities. The number of threatened species by different human activities are taken as a proxy for the threat to biodiversity. Endangered species probably play a minor role in ecosystem functions, so there is a weak causality between the red-list and ecosystem services. But there needs to be a correlation for the measure to be meaningful.

All values are adjusted to 2018 US$. US inflation rates are used.

The ISO standard 14008 has several requirements on what to report when determining a monetary value of an environmental good. Table 5.1 describes the methodological choices made here as required to be reported

Table 5.1. General characteristics of monetary values.

Method issue	Method choice
Environmental impact valued	Specified for each good
Increase or decrease valued	Decrease
Environmental impact indicator	Specified for each good
Indicator unit	Specified for each good
Quantity of environmental impact the value is for	One indicator unit
Environmental baseline	Business as usual
Monetary valuation method	Specific to each good
Justification of method choice	Specific to each good
Base year	2018
Currency	US$
Uncertainty	Specific to each good
Spatial coverage	Global
Temporal coverage	Specific to each good
Affected population	Global, intergenerational
People, whose values are determined	OECD population
Equity weighting	None
Marginal, average or median	Logarithmic average ≈ median when possible

by the standard. By default, global values are determined. Product systems normally involve processes distributed over large areas, which makes it reasonable to start an impact valuation with global averages. If a sensitivity analysis indicates that local impacts play an important role, local values can be determined using similar methodology.

Below, market or replacement costs are determined for environmental impacts on a number of environmental goods. Each clause starts with a description of the impact and good that is valued. Second, one or several methods are applied to estimate the value as a best estimate. Third, the uncertainty is assessed. Finally, a conclusion is made of which best estimate and uncertainty to use in the modelling of environmental impacts values from emissions and extraction of natural resources.

5.1 Production Capacity of Crops

The impact valued is decreased production capacity of crops. The impact indicator is decreased harvests. The indicator unit is kg crop. The monetary valuation method used is producer price.

A crop is "a cultivated plant that is grown on a large scale commercially, especially a cereal, fruit, or vegetable" (*Oxford Dictionary*). Production capacity is what is possible to sustainably harvest with the present agricultural methods. The absolute value of the production capacity is not necessary to know, we only need to be able to value changes in production capacity due to environmental impacts.

The major environmental impacts on the production capacity of crops come from climate change, oxidants, and land use. For crops, there are markets and market prices that can be used for the estimate of their value. Producer prices will be used. Producer prices reflect value losses due to environmental impacts better than market prices to consumers. This is because the cost of farming will be approximately the same even if the harvest decreases, while income depends on the harvest volume. Cost of distribution and marketing largely depend on volume and will not affect the crop producer price per kg.

The price of crops varies depending on which crop is produced, but it is here assumed that the present production mix mirrors present human needs. In Table 5.2 global production and producer prices are given for the major crops [1]. FAO reports annual production and average prices for most countries in the world. The weighted median crop price for the last ten-year period is 271 $/ton, which transferred to 2018 $ is 289 $/ton or

Table 5.2. World average producer prices for major crops for the years 2008–2017 [1].

Crop type	Median price ($/ton)	LN median price	Production (Ton)	LN price* production	Uncertainty in Median price (%)
Wheat	234	5.4553	5.74E+09	3.13E+10	1.6
Rice	381	5.9428	4.79E+09	2.85E+10	2.3
Maize	213	5.3613	8.58E+09	4.60E+10	2
Soybean	460	6.1312	2.39E+09	1.46E+10	2
SUM			2.15E+10	1.20E+11	
Average		5.7227			2.0
Standard deviation		0.3730			

0.289 $/kg. 1 kg is the indicator unit that will be used later in modelling. The uncertainty of the overall crop median value estimated from the standard deviation in the specific crop type prices is 24%. The calculations are made as follows:

- Uncertainty in median price for all crops = EXP(0.3730/SQRT(4-1)) = 1.24 or 24%
- LN for weighted average median price = 1.20E+11/2.15E+10 = 5.6014
- Weighted average median price = EXP(5.601467) = 271 $/ton
- Correction for inflation: 271*1.0677 = 289 $/ton

5.2 Production Capacity for Meat

The impact valued is decreased production capacity of meat. The impact indicator is decreased production. The indicator unit is kg meat. The monetary valuation method used is producer price.

Meat is here defined in terms of livestock weight. This is partly because statistics are on livestock weights and partly because of uncertainty on what parts that are used as food. As a rule of thumb 40–50% of an animal is what we normally associate with "meat".

The production capacity for meat is threatened by climate change and land use practices. Median producer prices for major livestock are shown in Table 5.3. The average livestock price for the period 2008–2017 was 2430 $/ton, and 2590 in 2018 $ and 2.59 $/kg. The uncertainty of the median is 32%.

Table 5.3. World median producer prices for livestock for the years 2008–2017.

Type of livestock	Price, $/ton
Cattle	2155
Chicken	1792
Pig	1826
Sheep	4929
Median	2428
Standard deviation, %	61
Uncertainty of median, %	32

5.3 Production Capacity for Fish

The impact valued is decreased production capacity of fish. The impact indicator is decreased production. The indicator unit is kg fish. The monetary valuation method used is producer price.

"Fish" here also includes shellfish and mollusks, harvested both from freshwater and seawater. Acidification, climate change and eutrophication have an impact on the production capacity for fish. Mostly, the impacts decrease production capacity, but addition of nutrients may increase it.

Fish prices are only available for traded fish, but these are assumed to be a good proxy for the value of locally produced and consumed fish [2]. Aquaculture's share in direct human consumption was 55% in 2017, according to FAO. For 2016, FAO estimates a total production of 60–60.5 million ton and sales at 141.6 billion $. This gives an average price of 2360 $/ton or 2420 transferred to 2018 $. The value per kg will be 2.42 $, which is notably close to the meat value. The uncertainty is similar as for meat, i.e., 30%.

5.4 Production Capacity of Wood

The impact valued is decreased production capacity of wood. The impact indicator is decreased production. The indicator unit is m^3 wood. The monetary valuation method used is producer price.

A tree contains wood in its stem, branches and roots. Stem wood has the highest commercial value and changes in production capacity for stem wood will determine the monetary value of an environmental impact. However, branches and roots are not without monetary value either as firewood, as storage for climate gases and as part of ecosystem processes, but there is not sufficient information to ascribe them a monetary value.

Prices of wood vary much due to varying use and cost for transports to users. FAO published data for export prices for roundwood from different countries [3]. The median or "weighted logarithmic average price" for 2017 was 82 $\$/m^3$ solid wood (excl. bark), which is equivalent to 84 \$ in 2018. The uncertainty of the median value is 7%, based on the standard deviation in prices from 37 different countries. Depending on the moisture content, there is an uncertainty of how much fibre there is in the wood. In Scandinavia "shipping dry" for loose goods means 20–24% moisture and for packed goods it is 18–19% [4]. Moisture in wood is defined as water relative to dry wood weight, i.e., it can be more than 100%. Here a 20% moisture content is assumed for traded roundwood. The density of wood mostly varies between 500 and 900 kg/m^3. 700 is used as an average.

Forests are not just producing roundwood for sawmills. They also produce pulp wood for paper and fuel wood. No general information on global basis has been found, but in Swedish boreal forests about 50% of the wood is used for pulp and fuel, selling at 60% of that of roundwood for sawmills. Applying that figure on the roundwood values results in a price of 72 $\$/m^3$. However, the uncertainty increases as temperate and tropical forests probably have different costs and volumes for pulp and fuelwood. Given that the price correction for Swedish boreal forests was 20%, the uncertainty in terms of standard deviation should be lower, but as the uncertainty for the average price of roundwood was 7%, a total of 20% seems reasonable.

The best estimate is thus 72 $\$/m^3$ with an uncertainty of 20%, which is a factor of 1.2.

5.5 Biodiversity

The impact valued is decreased biodiversity. The impact indicator is share of threat to red-listed species. The indicator unit dimensionless. The monetary valuation method used is preservation costs.

The quantitative value of the environmental goods made available through biodiversity is not possible to estimate at present. There is not enough knowledge about how biodiversity links to ecosystem services. Nor is there a consensus on how to measure biodiversity. The only way found to estimate its value is by looking at the cost of measures considered necessary for the preservation of biodiversity.

McCarthy et al. [5] estimated the total Financial Costs of Meeting Global Biodiversity Conservation Targets, set up by governments, to US\$ 76.1 billion per year. Thus, the best estimate of the total value of biodiversity is 76.1 billion \$/year. Changes in biodiversity from a single human activity is in this book valued according to the activity's share of threats to red-listed species. Threat causes for red-listed species are found in IUPCNs database.

The uncertainty in the estimate is assumed to be 50%.

5.6 Production Capacity of Drinking Water

The impact valued is decreased production capacity of drinking water. The impact indicator is decreased production. The indicator unit is m^3 water. The monetary valuation method used is producer price.

Drinking water is here defined as water that fulfils the criteria in WHO's "guidelines for drinking water quality". The production capacity is the flow of drinking water from a source. For most drinking water consumed, there is a treatment plant that adjusts the water quality to comply with the WHO guidelines. Such water will have a value decreased by the treatment cost.

Access to water of good quality is one of the largest problems in the world today. Local land use practices, climate change and water pollution decrease drinking water availability. The indicator valued here is not the use of water per se, it is the decrease of water production capacity in the environment. Value loss may be due to emptying stocks of fossil water, decreasing precipitation or evaporation or changing water flows. The indicator unit used here is 1 m^3 of drinking water.

Market prices for drinking water is available from IWA (International Water Association) [6]. The median value for the 28 countries for which values are available is 1.87 $ per m^3. This value is equity adjusted using country specific GDP with OECD average as reference. The standard deviation is a factor 2.26, which is quite high, but reflects local variations and the local character of water resource economy. For example, for crops world average and median values may be determined with high precision. For water, there is no global trade, and local prices vary, resulting in a relatively high uncertainty.

The best estimate of the value of production capacity for drinking water is thus 1.87 $/$m^3$ with an uncertainty factor of 2.26.

5.7 Coal Resources

The impact valued is decreased coal resources. The impact indicator is extraction of coal from present reserves. The indicator unit is kg coal. The monetary valuation method used is costs for substituting coal with a sustainable alternative.

Coal is a sedimentary rock, which is traded and used as fuel and raw material in industrial processes. Coal exists in many qualities. Anthracite has the highest coal content (86–97%), bituminous coal contains 45–86% carbon and subbituminous coal contains 35–45% carbon. Anthracite and bituminous coal are called hard coal. Subbituminous coal and lignite are called brown coal. Most coal is used for heating and called thermal coal. Certain bituminous coal with low ash content and low sulphur and phosphorous content is called metallurgical coal and are used to make coke for steelmaking. In many reports on the use of coal, it is not clear what type of coal is used. This introduces uncertainty in the monetary valuation of coal. There is however no point in being too specific about the value of different kinds of coal, as there is seldom knowledge on what type of coal that has been used in, e.g., energy production. The approach used here is to value as pure coal as possible, i.e., anthracite, and if known what other types of coal that is used, the value can be modified by the reduction in coal content.

Although a relative abundant resource, which may still allow mining for several hundred years, this is still a short time in a historical perspective and future generations will have to find a substitute. Not knowing all future uses of coal, the substitute chosen should be as similar to coal as possible. Charcoal is such a substitute, which is produced and sold today. One may argue that wood would be a substitute itself, when coal is used as fuel, but wood cannot substitute coal in all applications. Several studies have dealt with replacing coal with charcoal in steelmaking. There are a few studies in the literature reporting prices of charcoal as between 160 and 780 $/ton [7]. A recent market survey finds a price of 270 $/ton for 2013 [7]. This is equal to 284 $/ton 2018 or 0.284 $/kg. Price variations depend much on varying prices of wood. For wood, an uncertainty of 20% was assumed for the median value. Considering that there also is an uncertainty about the value of co-products from the pyrolysis process, the uncertainty should be higher. A 30% uncertainty is assumed.

5.8 Lignite Resources

The impact valued is decreased lignite resources. The impact indicator is extraction of lignite from present reserves. The indicator unit is kg lignite. The monetary valuation method used is costs for substituting lignite with a sustainable alternative.

As mentioned for coal, the classification and terminology vary between different countries and publications. In Merriam-Webster's dictionary lignite is defined as "a usually brownish black coal intermediate between peat and bituminous coal *especially*: one in which the texture of the original wood is distinct". The lignite that is valued here has a high content of water (about 50%), its coal content is 60–70% on a dry, ash-free basis. Its calorific value is 14 MJ/kg in moist condition.

A substitute for lignite could be peat or wood. Peat is a slowly renewable resource and may only replace lignite in low volumes. There is however a market for peat, and its price is round 30 $/ton.

Lignite can be substituted by wood based on its energy content. The coal content of wood is somewhat lower than in lignite, about 50% [8]. The net calorific value for lignite is 14 MJ/kg and 16 MJ/kg for air-dry wood. Wood taken out of the forests often have a dry substance(DS) content of 55%, while air-dry wood has about 70–75% DS. Wood taken out of the forests has a density close to 1000 kg/m^3. Freshly cut logs barely float. Real dry wood with a few % moisture has about half the density, 500 kg/m^3. There are more dense wood types, but 500 is relevant for many common tree species, such as pine, spruce and birch. Air dry wood is estimated to have a density of 750 kg/m^3. Its value was estimated in 5.4 to 72 \$/m^3. This means that the value of 1 kg of lignite would be 72/750*14/16 = 0.084 \$/kg. The uncertainty is similar as for wood, i.e., 20%.

5.9 Oil Resources

The impact valued is decreased oil resources. The impact indicator is extraction of oil from present reserves. The indicator unit is kg oil. The monetary valuation method used is costs for substituting oil with a sustainable alternative.

Like for coal, crude oil is available in a variety of qualities. According to IEA (International Energy Agency), "Crude oil is a mineral oil consisting of a mixture of hydrocarbons of natural origin and associated impurities, such as sulphur".

A substitute for oil with an almost identical chemical composition is HVO diesel oil. HVO diesel is beginning to be produced in larger quantities and reveal a market price. It is presently (January 2019) sold in Sweden for 1.80 \$/liter. As energy and carbon dioxide taxes is refunded, the tax-free market price can be estimated to be 0.88 \$/liter or 1.06 \$/kg. An alternative would be to use a Fisher-Tropsch process with wood as an input. It would lower the cost to 0.5 \$/kg [9], but the chemical composition would be different. The uncertainty in estimating the cost for a substitute to fossil oil is much due to the fact that oil is not a very well defined good. To include this broad variation a best estimate of 0.8 \$/kg is assumed with an uncertainty of 30%.

5.10 Natural Gas Resources

The impact valued is decreased natural gas resources. The impact indicator is extraction of natural gas from present reserves. The indicator unit is kg natural gas. The monetary valuation method used is costs for substituting natural gas with a sustainable alternative.

According to IEA, "Natural gas comprises gases, occurring in underground deposits, whether liquefied or gaseous, consisting mainly of methane."

Methane may be produced as biogas by anaerobic digestion. The production cost for biogas from anaerobic digestion of animal manure is estimated to be between 2.99 and 6.82 2006 \$/GJ with an average of 6 \$/GJ for 17 digesters [10]. Natural gas has a low heating value of around 48 MJ/kg. This means that production cost for a natural gas equivalent biogas would be 6*0.048 = 0.288 \$/kg or 0.36 2018 \$/kg. If there is a cost for the biomass, e.g., if wood is used with a cost of 72 \$/m^3, there is an additional 0.27 \$/kg. However, it is not likely, that one needs to use high quality wood for biomass and that there are cheaper solutions. A best estimate of 0.5 \$/kg is therefore assumed with an uncertainty of 30%.

5.11 Aluminum Resources

The impact valued is decreased aluminum resources. The impact indicator is extraction of aluminum from present bauxite reserves. The indicator unit is bauxite containing 1 kg natural aluminum. The monetary valuation method used is costs for substituting bauxite with a sustainable alternative.

The main natural resource for aluminum is bauxite. Bauxite is a sedimentary rock with a relatively high aluminum content (25–55%).

Aluminum is one of the most abundant elements in the earth's crust. The average concentration in the upper continental crust is 7.14% [11]. Several substitutes have been proposed for bauxite as a source of aluminum, like clay minerals, alunite, anorthosite, power plant ash, and oil shale.

During the Second World War, aluminum was produced from andalusite (Al_2OSiO_4) in Sweden. The mineral was leached by sodium hydroxide to produce aluminum oxide which was used in a similar way as the NaOH leachate form bauxite [12]. The cost of replacing bauxite as a raw material, is assumed to be mainly caused by the extra material that must be mined, crushed and grinded. When extracting $Al(OH)_4^{-1}$ from the milled stone by use of NaOH, it is mainly Al_2O_3 that reacts. SiO_2 could also react, but only at very high temperatures (900–1000°C).

The cost of open pit mining is estimated to be 3.1 \$/ton total material [13]. The cost of crushing and grinding was given for magnetite ore by Jankovic and Valery [14]. Adjusted to 2018 \$, the cost of crushing and grinding is 6.7 \$/ton. Ordinary rock, like granite, takes a little more work to crush and grind than magnetite. The work index of granite is 14.4 kWh/ton compared to 10.2 kWh/ton for magnetite [15]. Adjusted for the extra work, the cost is 9.4 \$/ton. The total cost for producing stone fines is 12.5 \$/ton. With an average Al concentration of 7.14%, this is 175 \$/ton Al or 0.175 \$/kg Al.

There are two major factors contributing to the uncertainty in the cost estimate. One is the concentration of Al in the minerals that will be substituting bauxite. There are probably large amounts of minerals with higher Al content than 7.14% that can be used "forever". Another uncertainty is the consumption of NaOH, when extracting Al from minerals. The stochiometric consumption is shown in the formula below:

$$Al_2O_3 + 2NaOH + 3H_2O \rightarrow 2\,Al(OH)_4^- + 2Na^+$$

NaOH costs about 125 $/ton and at stochiometric amounts its cost is 185 $/ton Al. It is likely that some extra loss of NaOH occurs, when the amount of material processed increases, but this uncertainty counteracts the first uncertainty. Therefore, an uncertainty of 30% is assumed, mainly because of uncertainty of the cost for NaOH.

In earlier versions of cost calculations made for use in the EPS system, external costs of emissions and resource use were included, but as their contributions were small compared to the uncertainty in ordinary tangible cost estimates [16], they are left out this time.

5.12 Iron Ore

The impact valued is decreased Fe resources. The impact indicator is extraction of iron from present iron ore reserves. The indicator unit is iron ore containing 1 kg Fe. The monetary valuation method used is costs for substituting iron ore with a sustainable alternative.

Iron ore consist of various kinds of oxides. Many ores contain more than 60% Fe and can be fed directly into blast furnaces for ironmaking. But ores with down to 15% Fe is also mined today.

Iron is by far the most used of all metals. The global production of crude steel was 1689 million tons in 2017. The second most used metal, Al, was produced as primary metal in 60 million tons. This means that to replace such volumes, Fe must be mined separately. The average concentration of iron in earth's continental upper crust is 3.91% [11], but there are significant variations between granite, granodiorite and basalt, which makes up about 85% of the crust mass. The global average Fe concentrations in common igneous rocks are shown in Table 5.4.

A sustainable process for producing iron ore substitutes would start with producing stone fines for extraction with HCl. HCl has been found to be the most feasible acid. Extraction efficiencies of 86.1, 80.1 and 44.6%

Table 5.4. Average concentrations in common igneous rock types [17].

Substance	Granite	Granodiorite	Diorite	Syenite	Anortosite	Gabbro	Basalt
Fe_2O_3	1.22	1.52	2.55	3.05	0.99	3.1	3.9
FeO	1.68	2.87	5.13	3.22	2.07	7.76	7.5
Fe	2.16	3.29	5.77	4.64	2.30	8.20	8.56

was found for granite, granodiorite and basalt, respectively, when using 30% HCl [18]. The consumption of HCl was roughly corresponding to the extracted metals and determined to 47, 56 and 88 kg pure HCl per ton granite, granodiorite and basalt fines. After leaching the leachate has to be neutralized with NaOH to precipitate Fe as $Fe(OH)_2$ or $Fe(OH)_3$.

The cost for producing one ton of Fe ore substituted in this way, is estimated in Table 5.5.

A critical issue in estimating substitution costs, is probably whether you need hydrometallurgical methods or not. If the same magnetic concentration methods can be used as is used today, after mining, crushing and grinding, the cost of producing substitutes would roughly be proportional to the inverse of the Fe concentration. In the mining today, ore grades vary between 20 and 60% Fe. If the lowest grade represents what is economically feasible, the present world market price may be used as an estimate of the production cost to produce an ore with 62% Fe from a 20% ore. The market price today is 70 $/ton. The last five years it has varied between 60 and 140 $/ton as a yearly average. 70 $/ton is assumed to be what a mining company would need to survive. If 70 $/ton Fe is the production cost to achieve 62% iron from 20% ore, then the cost per ton of processed material is 70*20/62 = 22.5 $/ton. Producing 62% ore from crustal average of 3.91% would cost 70*3.91/62 = 359 $/ton or 579 $/ton Fe. For granite, granodiorite and basalt, the cost estimates would be 1046, 686 and 264 $/ton Fe. However, it seems likely that the Fe mineral crystals require milling to smaller particle sizes to be efficiently separated from non-Fe minerals. The cost estimates are therefore considered to be a lower bound.

The uncertainty in the cost estimate for the hydrometallurgical process is assumed to be 30%, partly due to uncertainties in the cost for a sustainable production of NaOH and HCl from NaCl and partly due to uncertainties in which Fe concentration that can be sustainably mined.

The uncertainty of the cost for sustainably producing NaOH and HCl very much depends on how the electricity is produced. The costs of solar, wind and water power are today close to the cost of fossil power, why the uncertainty should be in the order of 20%.

The uncertainty in the sustainable concentration of Fe ore is mainly impacting on the volume of material subject to crushing and grinding. However, this cost is only about 20% of the total ore substitution cost (Table 5.5).

The largest uncertainty comes from not knowing which kind of process, that may work. The figure 0.26 $/kg Fe for enriching Fe from basalt via a metallurgical process is probably too low. The work index for grinding basalt is twice the work index for grinding magnetite, and there is probably a need to grind to smaller particle sizes. This means that the

Table 5.5. Cost calculation for hydrometallurgical production of iron ore substitutes.

Granite						Granodiorite		Basalt	
Unit process (UP)	**UP reference flow**	**UP reference flow unit**	**Cost per flow unit ($)**	**Flow units per ton of Fe**	**Cost per ton of Fe extracted ($)**	**Flow units per ton of Fe**	**Cost per ton of Fe extracted ($)**	**Flow units per ton of Fe**	**Cost per ton of Fe extracted ($)**
Mining, crushing, grinding	stone fines	ton	12.5	53.8	672.4	37.9	473.8	26.2	434.7
Leaching	33% HCl acid	ton	150	7.6	1137.7	6.4	963.3	6.9	1037.6
Precipitation	NaOH	ton	125	8.4	1053.4	7.1	892.0	7.7	960.7
Waste management	leaching water	ton	0.32	10.0	3.2	10.0	3.2	10.0	3.2
Waste management	fines	ton	1	53.8	53.8	37.9	37.9	26.2	26.2
				Sum	2920		2370		2462

cost lies between 0.5 \$/kg Fe for a metallurgical process and 2.37 \$/kg Fe for a hydro-metallurgical process. A best estimate of 1.1 \$/kg Fe with an uncertainty factor of 1.6 will therefore be used. The high and low values will then be within two standard deviations in a log-normal distribution.

5.13 Base and Trace Metal Ores (Except Alkali Metals)

The impact valued is decreased metal resources. The impact indicator is extraction of a metal from its present ore reserves. The indicator unit is ore containing 1 kg of the metal. The monetary valuation method used is cost for substituting the metal ore with a sustainable alternative.

Most base and trace metal ores contain several metals, which are separated by processes which are adapted to their composition. It would be too complicated here to value specific ore mixtures and why each metal concentrate is valued separately, e.g., per kg Cu in ore.

While Al and Fe can be sustainably mined separately, because of the large quantities needed, and high concentrations available, other metals are preferably mined together and processed by hydrometallurgical processes. There is not much known about how exactly this can be done, but I have made tests together with a colleague on leaching granite, granodiorite and basalt [18]. We used standard samples of basalt, granite and granodiorite, which were purchased from the US Geological Survey. The standard reference material samples were ground and were accompanied by certificates showing the results of performed chemical analyses. We tested different leaching agents and found HCl to be the most efficient leaching agent in terms of average and median leaching efficiency, and the potentially most sustainable alternative. NaCl is available in large quantities and can be processed by electrolysis to yield HCl and NaOH. The leaching efficiencies we obtained are shown in Table 5.6.

Other metals may need other leaching agents. NaOH, H_2S and some sequestrants and organic solvents may be used to precipitate or extract the metals from the leachate. In some cases, it may be necessary to evaporate the solvent to obtain sufficient concentrations. As a first approximation, the cost of producing an ore-like concentrate of a metal is assumed to be inversely proportional to its concentration in the rock times its leaching efficiency.

The cost would then be $K/(C_i{}^*\eta_i)$, where K is a constant, representing the cost of chemicals and energy for the hydrometallurgical process, C_i, the concentration of the metal i in the rock and η_i the leaching efficiency for metal i. The cost for hydrometallurgical production of Fe-ore substitute from granodiorite was 2370 \$/ton Fe. The Fe concentration was 30335 g Fe/ton rock and the leaching efficiency 0.801, which would give a $K = 57.6$ \$/ton rock. In reality, several metals are likely to be mined at the same

Table 5.6. Extraction efficiency of some metals when leaching in 30% HCl for 24 hours and the first hour heated to 50°C.

	Granite		Granodiorite		Basalt	
	Content (g/ton)	**Mean yield (%)**	**Content (g/ton)**	**Mean yield (%)**	**Content (g/ton)**	**Mean yield (%)**
Al	81497	9.9	80703	14.5	72182	5.6
Cr	7	100.8	12.5	62	17.6	10.2
Fe	18824	86.1	30335	80.1	94139	44.6
Mn	260	90.9	331	81.8	1406	27.8
Co	5.5	73.3	6.4	76.9	38	35.5
Ni	5.1	44.9	12.5	51	15.8	24.2
Cu	11.7	79.9	33.3	80.2	18.4	(126.6)
Zn	85	90.5	98	85.1	120	62.9
Sr	479	5.5	233	8.5	330	5.5
Cd	0	22.5	0.1	50	0.1	33.7
Ba	1870	14	1300	11.3	675	2.7
Pb	31.2	22.1	51.3	65.3	17.6	38.4
Sn	1.5	76.2	6.3	72.9	2.6	58.3
W	0.2	27.6	0.3	22.3	0.4	63.4

time. 20 are assumed as a default. Then K will have a value of 2.88 $/ton. Table 5.7 shows the values obtained for trace metals at average crustal concentration with $K = 2.88$ $/ton. If no leaching efficiency is empirically determined, a 50% efficiency is assumed.

The uncertainty for the cost estimate for Fe was set to 1.6, but the uncertainty in estimation of the cost for trace metal ores is larger because we do not know how many metals will be mined at the same time. If only ten metals are mined, then the cost will be doubled and if all 54 metals in Table 5.7 are mined at the same time, the cost would be less than half. An uncertainty of a factor 2 is therefore assumed.

5.14 Alkali Metal Resources

The impact valued is decreased alkali metal resources. The impact indicator is extraction of an alkali metal from its present reserves. The indicator unit is concentrates containing 1 kg of the alkali metal. The monetary valuation method used is cost for substituting the alkali metal concentrate with a sustainable alternative.

Salt deposits and sea water are the main sources of Na and K. Salt deposits and minerals are also sources of **lithium**, Li, **rubidium**, Rb and **cesium**, Cs. Li are mined from several types of minerals containing a few % of Li and from salt deposits containing between 600 and 1500 ppm when

Table 5.7. Estimated cost of production of trace metal ore substitutes.

Metal name	Abundance in crust (g/ton)	Cost for ore substitution ($/kg Metal)	Metal name	Abundance in crust (g/ton)	Cost for ore substitution ($/kg Metal)
Silver	0.05	115200	Neodymium	26	222
Arsenic	1.5	3840	Nickel	44	136
Gold	0.0018	3200000	Osmium	0.00005	115200000
Bismuth	0.13	44308	Lead	17	339
Cadmium	0.098	58776	Palladium	0.00053	10867925
Cerium	64	90	Praseodymium	7.1	811
Cobalt	17	226	Platinum	0.0006	9600000
Chromium	83	43	Rhenium	0.0004	14400000
Cesium	4.6	1252	Rhodium	0.000018	320000000
Copper	25	144	Ruthenium	0.00003	192000000
Dysprosium	3.5	1646	Antimony	0.2	28800
Erbium	2.3	2504	Scandium	14	411
Europium	0.88	6545	Samarium	4.5	1280
Gallium	17	339	Tin	5.5	698
Gadolinium	3.8	1516	Tantalum	1	5760
Germanium	1.6	3600	Terbium	0.64	9000
Hafnium	5.8	993	Tellurium	0.001	5760000
Mercury	0.067	85970	Thorium	2.8	2057
Holmium	0.8	7200	Titanium	4100	1
Indium	0.05	115200	Thallium	1	5760
Iridium	0.000022	261818182	Thulium	0.33	17455
Lanthanum	30	192	Uranium	10.7	538
Lithium	20	288	Tungsten	2	5760
Luthetium	0.32	18000	Vanadium	107	54
Manganese	600	5.6	Yttrium	22	262
Molybdenum	1.5	3840	Ytterbium	2.2	2618
Niob	12	480	Zinc	71	46
			Zirconium	190	30

extracted as brine [19]. Rubidium is mined from Lepidolite, which contains between 0.3% and 3.5% Rb. Cs is mined in the form of the mineral pollucite.

All alkali metals may be sustainably obtained from seawater. Many attempts to extract Li from seawater have been published [20]. Some investigate different sorbents and some membranes for ion separation. Shahmansouri et al. [21] estimated the potential to economically extract different minerals from run-off concentrates of desalination plants. They estimated the economic potential from the expression

$$P \times C \times Q_c - OM > 0,$$

where, P is the market price of the element, C is its concentration in the run-off brine, Q_c is the flow of the run-off brine and OM is the operation and management cost. No capital costs are included in their calculations. If $P \times C \times Q_c - OM<0$, $-(P \times C \times Qc - OM)$ can be used as a lower estimate of the replacement cost of a mineral. Such estimates are made in Table 5.8 based on data from Shahmansouri et al. [21].

If Li and Cs were obtained from earths average crust as for trace metals, the costs would be 180 and 780 $/kg respectively.

The uncertainty in the assessments is high. Although potentially sustainable extraction processes can be identified, their efficiencies are largely unknown. An alternative way of obtaining Li from sea water is to bring large amounts of water into basins where the water is evaporated by sunlight. Current production of sea salt is made in this way. Because of different solubility (Table 5.9), Li ions will remain and concentrate in the brine when Na and K crystalize. Today most of the residue brine, contain Mg, has a bitter taste and is discarded. Using such brine would allow lower costs for Li minerals, not much different from present production costs. But there may be problems of producing sufficient volumes. Only a small part of the NaCl salt produced today is made via solar induced evaporation. The global NaCl production 2017 was 280 million ton. In the US sea salt was only 9% of the NaCl production. If all 280 million tons of NaCl would be produced by solar evaporation of sea salt, 5600 ton Li could be obtained per year. A slightly higher amount (6000 ton) could be obtained from desalination plants operating today. According to Choubey et al. [24] there is a need for 93 000 ton Li 2025 if current trends on electrification of vehicles are projected. In a scenario with only BEV cars and 80% recycling of the batteries, the need for Li is likely to stabilize around 500 000 ton/year [25]. This means that a sustainable production of Li from seawater to support the battery industry is challenging as the

Table 5.8. Estimated replacement costs for alkali elements as ions.

Element	Concentration in sea water (g/m^3)	Replacement cost ($/kg)
Li	0.18 [22]	10
Na	10800	0
K	400	0
Rb	0.12 [23]	0
Cs	0.0003	30000

Table 5.9. Solubility of various ions of Li, Na, K and Mg.

Substance	Solubility (g/l)	Water temp. (°C)
LiCl	459	25
Li_2CO_3	7.2	100
Li_3PO_4	< 0.34	
Li_2SO_4	257	20
NaCl	264	20
Na_2CO_3	485	100
Na_3PO_4	45	
Na_2SO_4	71	20
KCl	258	25
K_2CO_3	1560	100
K_3PO_4	1930	
K_2SO_4	107.5	25
$MgCl_2$	164	20
$MgCO_3$	0.1	100
$MgSO_4$	225	25

recycling efficiency then would need to be around 99.6% if 10 000 tons of virgin Li would be enough as a make up of the Li stock in use.

To conclude:

For Na, K and Rb, the best estimate of the replacement cost is zero.

For Li it is between zero and 10 $/kg. The best estimate chosen is 5 $/kg with an uncertainty of a factor of 2.

For Cs, the best estimate is 780 $/kg, with an uncertainty of a factor of 2.

5.15 Alkali Earth Metals

The impact valued is decreased alkali earth metal resources. The impact indicator is extraction of an alkali earth metal from its present reserves. The indicator unit is minerals containing 1 kg of the alkali earth metal. The monetary valuation method used is costs for substituting the alkali earth mineral with a sustainable alternative.

Beryllium, Be is produced commercially from the minerals beryl and bertrandite. Beryl is found in granite and other igneous rocks. Unlike beryl, which is easy to distinguish from its surrounding minerals in the rock, bertrandite mining involves frequent drilling, sampling and analyzing [26]. Almost all Be is obtained from bertrandite.

No sustainable process for producing Be-ore-like substances are known. The abundance of Be in the earth's crust is 3 mg/kg. It is claimed to have similar chemical properties as Al. The leaching efficiency of Al from granite was 9.9% in HCl (Table 5.6). Using the same formula as for base and trace metal ores the substitution cost will be 9600 $/kg.

The uncertainty depends largely on the leaching efficiency estimate. Maybe Be can be co-mined with Al. The uncertainty is assumed to be a factor of 5.

Main sources of **magnesium**, Mg are dolomite, magnesite and sea salt. Mg is the 8th most abundant element in the earth's crust. The extraction of Mg is near sustainable. The cost for a sustainable substitution of Mg is thus set to 0.

Calcium, Ca is the fifth most abundant element in the earth's crust. Ca can be obtained from limestone deposits, and the process is close to sustainable. The substitution cost is therefore set to 0.

Strontium, Sr is the 15th most abundant element in the earth's crust. It is mined from the minerals strontianite (strontium carbonate) and celestite (strontium sulfate). Celestite is the main commercial source for Sr. Strontium may also be obtained from desalination of sea water. The potential cost is then in the order of 0.2 $/kg [21]. The uncertainty depends on the flows treated in the desalination plant and is estimated to a factor of 2.

Barium, Ba is the 14th most abundant element in the earth's crust. Its commercial source is Barite. In our leaching experiments we obtained 14% leaching efficiency from granite with a Ba concentration of 1870 mg/ton (table y). This indicates a sustainable production cost of 11 $/kg Ba. The uncertainty is believed to be similar to that of Sr, a factor of 2.

Radium, Ra is a very rare element and part of the decay chain of uranium. This makes the issue of sustainable mining somewhat irrelevant, and that is why its resource value will not be assessed here.

5.16 Non-metal Resources Excluding Halides

The impact valued is decreased non-metal resources. The impact indicator is extraction of a non-metal element from its present reserves. The indicator unit is concentrates containing 1 kg of the non-metal element. The monetary valuation method used is the cost for substituting the non-metal concentrate with a sustainable alternative.

Most of the **hydrogen** produced today comes from natural gas. There is also a sustainable alternative in electrolyzing water. The cost of hydrogen from steam reforming of natural gas or other fossil resource is around 0.75 $/kg [27] or 1.5 $/kg if CO_2 costs are included, either as external costs or as sequestration costs. These cost estimates were from 2014. A more recent cost estimate from 2018 reports costs of 2.1 $/kg hydrogen including carbon sequestration [28]. Steam reforming is an established technology and costs are fairly stable.

Hydrogen may be produced sustainably by biomass gasification or biomass pyrolysis or by electrolysis. The cost of hydrogen from biomass gasification is estimated to be 1.21–2.42 $/kg and hydrogen from biomass pyrolysis to be 1.21–2.19 $/kg as of 2014. The cost estimate of hydrogen from electrolysis varies a lot with time and ways of calculation and what is included. The technical development is intensive, the cost of electricity can vary a lot and distribution costs are high. Some estimates use marginal electricity cost at the production site, which can be close to zero, and some use average grid costs. Compression and liquefaction can cost as much as 2 $/kg, the valuation here is based on normal pressures and temperatures. Parthasarathy and Narayanan [27] estimate the cost of producing hydrogen by electrolysis to 2.56–2.97 $/kg.

This means that the extra cost for substituting fossil-based hydrogen with a sustainable alternative is between 0 and 0.87 $/kg. The average of 0.44 $/kg is used as a best estimate and the uncertainty is set to a factor of 2.

Boron, B, is produced from the minerals colemanite, rasorite (kernite), ulexite and tincal. Together these constitute 90% of mined boron-containing ore. The average B concentration in the earth's upper continental crust is 17 mg/kg [11]. The average concentration of H_3BO_3 in sea water is 22 mg/l [29], which corresponds to 3.8 mg B/l. Absorption of boron from sea water by porous condensate cation exchangers has been demonstrated [30]. Efficiency and selectivity were good. The cation exchanger was regenerated with 10% sulfuric acid. A typical price for cation exchangers is between 1000 and 10000 $/m^3. If the performance reported for boron lasts for a year, about 5 kg boron may be separated per l of resin, giving a capital cost of 0.2–2 $/kg. Operation and management cost can be moderate if the plant is run as part of a desalination plant. With sulfuric acid cost of 300 $/ton and a sulfuric acid consumption corresponding to the eluted borate, that cost should be around 6 $/kg boron. Adding part of the operation and management costs a best estimate would be 10 $/kg with an uncertainty factor of 2.

For **carbon**, C, see clause on coal.

Silicon, Si, is the second most abundant element in the continental crust. Resources are so large that present mining is considered close to sustainable, and the extra cost set to 0.

Nitrogen, N, is obtained from N_2 in air in a potential sustainable way. The extra cost for sustainable substitution is set to 0.

Phosphorous, P, is a common subject for sustainability concern. Today most P is obtained from phosphate rock, containing 4–20% of P_2O_5 [31] and used as phosphate fertilizers in agriculture. Much can be done to improve sustainable P fertilizing, like P recycling, making P more accessible to plants, but if something similar to phosphate rock is needed, P minerals in ordinary rock is probably a sustainable solution.

The average crustal concentration of P is 0.7 g/kg. A 25-fold concentration would produce a rock with 4% P_2O_5. Most P in rocks is present as phosphates. A strong acid is able to react with the phosphates, like calcium phosphate to form phosphoric acid. Dorozhkin[32] made a review of literature on the dissolution mechanism of calcium apatites in acids. If made together with leaching for trace metals the cost would be $K/(C_i{}^*\eta_i)$, where K is 2.88 \$/ton rock, C_i 0.7 kg P/ton and the leaching efficiency assumed to be 0.5, i.e., 8.2 \$/kg P, or if mined separately 164 \$/kg P. Separate mining would not be necessary if production of present P fertilizers at present volumes are needed, but given its high cost it does not seem to be a likely scenario.

There is some uncertainty with regard to the leaching efficiency and the P separation process from the leachate, so the best estimate for the cost is set to 8.2 \$/kg P with an uncertainty factor of 2.

Oxygen, O, is obtained from O_2 in the air in a potential sustainable way. The extra cost for sustainable substitution is set to 0.

Sulfur, S, is presently obtained from desulfurization of petroleum products. Elemental sulfur can be obtained from volcanic activities. Sulfur is also present in sulfide ores and as sulfates in minerals and sea water.

Being a very abundant element, with an average concentration of 0.905 kg/ton in the continental crust and 0.53 kg/m^3 in sea water, elemental S from volcanic activities and sea sulfates are sustainable alternatives to S in fossil minerals. As there is very little processing to obtain volcanic S and sea salt S, its cost is set to 0.2 \$/kg with an uncertainty factor of 2. Before desulfurization of petroleum products, the price of S was between 100 and 200 \$/ton in 1998 US \$ value, which is 155–310 \$ in 2018 according to USGS statistics.

Selenium, Se, is obtained from sulfide ores, where Se replaces S in small amounts. Se/S ratios are in the range of 1: (5000–10000). Se is a well distributed element in the earth's crust and the concentrations in sulfide ores are not dramatically higher [33–35] than the average concentrations in the continental crust 90 ppm [11]. Se concentrations in sulfide ores seem to be in the range 100–1000 ppm. Much of the value of sulfide ores as resources for Se seems to be in the co-mining with metals and sulfur.

A sustainable source of Se could be river, rain or sea water, where Se is present in concentrations between 0.004 and a few mg/m^3 [36]. Another option is to recover it from leachate of trace metals. In that case the cost will be 64 000 $/kg. A third option is to use bioaccumulation. Se is an essential element and may be accumulated in plants to as much as 3 mg/kg dry weight [37]. This is more than in sulfide ores. If the value of the plant biomass can be preserved by, e.g., energy production, the extra cost for Se production compared to that from sulfide ores, would be low.

The cost is assumed to be between 0 and 100 $/kg with a best estimate of 50 $/kg and an uncertainty factor of 5.

5.17 Halides

The impact valued is decreased halide element resources. The impact indicator is extraction of a halide from its present reserves. The indicator unit is concentrates containing 1 kg of the halide element. The monetary valuation method used is cost for substituting the halide concentrate with a sustainable alternative.

The main source of **fluorine**, F, today is fluorite, or fluorspar, CaF_2. Fluorspar is labeled as a critical resource by EU.

The average concentration in the continental crust is 557 mg/kg. F is mostly binding to Ca and it should be possible to separate it as gaseous HF after treatment with H_2SO_4. For producing 1 kg of F as HF, 2.68 kg of 96% H_2SO_4 is needed on stoichiometric basis. Assuming a 50% efficiency and that co-production of other substances that react with H_2SO_4 or are leached covers the cost for H_2SO_4 consumption not related to HF, the H_2SO_4 cost for 1 kg of F as HF will be 2.68/0.5*0.2 = 1.07 $, where 0.2 is the price in $/kg for H_2SO_4. The cost for mining, crushing and grinding was earlier (Table 5.5) estimated to 12.5 $/ton fines. If only F were produced, the cost per kg F would be 12.5/0.557 = 22.44 $. As our earlier leaching test showed, H_2SO_4 was not as good a leaching agent as HCl, the number of elements or compounds extracted would probably be less. 5 it is assumed. Then the cost would be 22.44/5 + 1.07 = 5.56 $/kg. Capital and operating costs add to that, but the process is not well defined, so no attempt will be made here to specify it.

Considering the above calculations, the best estimate is set to 6 $/kg F, with an uncertainty of a factor of 2. The uncertainty depends mostly on the extraction efficiency and number of other elements or compounds produced at the same time.

Chlorine, Cl, is obtained from a sustainable resource, sea water, why its extra cost is set to 0 $.

Bromine, Br, is today extracted from salt lakes. It is potentially profitable to extract from desalination plants. Its resource value is set to 0 $.

The main source of **iodine**, I, today is brine, but earlier it was obtained from kelp.

Kelp may be regarded as a sustainable resource, but depending on the volume, desalination plants may be an alternative. From estimations made by Shahmansouri et al. [21], the cost of extracting Iodine is between 0 and 350 $, depending on the size of the plant. Considering the present market price of Iodine being around 30 $/kg and that kelp was earlier profitable, the cost for sustainable substitution is set to 30 $/kg I with an uncertainty of a factor of 2.

5.18 Noble Gases

The impact valued is decreased noble gas resources. The impact indicator is extraction of a noble gas from its present reserves. The indicator unit is concentrates containing 1 kg of the noble gas. The monetary valuation method used is cost for substituting the noble gas concentrate with a sustainable alternative.

Helium, He, is the second most abundant element in universe, but rare on earth. The atmosphere holds only 5 ppm. Helium is formed from radioactive decay of mainly uranium and thorium, diffuses to the atmosphere and further into space. Some He is trapped under sedimentary rocks together with natural gas. Natural gas may contain several % of He and is today the main source of commercial production.

It does not seem possible to replace current He production with a sustainable one. The current total He flux through the continental crust surface is 1.79 kton/yr [38] while the production from natural gas is 20 kton/yr [39]. At much lower rates, the present reserves in natural gas may be naturally refilled and sustainably extracted at about the same cost as today. A best estimate for the extra replacement cost is set to 0.

Neon, Ne, **Argon**, Ar, **Krypton**, Kr, and **Xenon**, Xe, are all produced from cryogenic distillation of air, and if sustainable energy and proper recycling of materials are made, the extra cost would be close to zero. So, a best estimate for these elements are set to 0 $/kg.

5.19 Human Health—Working Capacity

The impact valued is decreased human working capacity due to environmental conditions. The impact indicator is decreased performance during 16 wake hours per day. The indicator unit is person-hours 100% lost working capacity. A 10% lost working capacity for one hour is thus 0.1 person-hour lost working capacity. The monetary valuation method used is cost of labor.

As discussed in the chapter on goal and scope, the labor market in OECD countries are used for estimating the value of working capacity impacts. The population weighted average is 60000 $/person-year for 2018 [40]. The average working hours for OECD was 1744 hours/year in 2017. If including time necessary for commuting and preparing for work, the time allocated to work is around 2000 hours per year, corresponding to an income of 30 $/hour. This value is also used for loss of working capacity for private use.

The standard deviation among country- specific average salaries is 17000 $. The best estimate for working capacity will therefore be set to 30 $/hour, with an uncertainty of 1.25.

5.20 Human Health—Years of Lost Life Expectancy

The impact valued is decreased human working capacity due to lost life expectancy. The impact indicator is decreased working capacity. The indicator unit is person-years lost life expectancy. The monetary valuation method used is cost for labor.

Labor is valued at 60000 $/year. In OECD the average working hours are 1744 per year. If it includes time necessary for commuting and preparing for work, the time allocated to work is around 2000 hours per year, corresponding to an income of 30 $/hour. For an average person in productive age there is about 16 hours of activities per day, or 5840 hours per year. The average global life expectancy is 72 years and if the productive years are estimated to be 44, the average value of one year's activities is 5840*30*44/72 = 107067 $, which also is the value of 1 year of lost life expectancy (YLL).

There is an uncertainty of how many active hours there are per day and the productive years in life. A factor of 1.3 is assumed.

5.21 Human Health—Disabilities

The impact valued is decreased human working capacity due to disabilities. The impact indicator is decreased working capacity. The indicator unit is person-years lost working capacity. The monetary valuation method used is costs for labor.

To guide health care policy, the DALY concept has been developed and is used by WHO. DALY stands for Disability-Adjusted Life Years and can be thought of as one lost year of healthy life. DALY is the sum of YLL and YLD, where YLD is years lived with disability. To be able to sum YLL and YLD weighting factors have been determined to assess the health reduction of a disease and its sequalae (~symptoms). A weighting factor of 1 means that the health reduction is equal to death, and a factor of 0,

Table 5.10. DALY weighing factors for different disutility categories.

Disutility category	Unit	Weighting factor	Value ($)	Note	Uncertainty factor
Years of life lost (YLL)	Person years		107067	Working capacity lost due to premature death.	1.3
Undernutrition	Person years	0.06	6424	Wasting protein-energy undernutrition with some development disability	1.1
Diarrhea	Person years	0.202	11242	Caused by polluted drinking water	1.5
Malaria episodes	Person years	0.191	20450	Potential impact from climate change	1.1
Gravation of angina pectoris	Person years	0.06	6424	Average for gravation of mild to severe symptoms	1.5
Cardiovascular disease	Person years	0.1	10707	Median value for several categories	2.4
Infarcts	Person years	0.0804	8795	Acute myocardial infarction: days 1–2, 0.422 (0.28–0.566) + days 3–28, 0.056 (0.035–0.082)	1.3
Asthma cases	Person years	0.043	4604	Treated asthma. Untreated 0.054	2
Chronic obstructive pulmonary disease, mild and moderate	Person years	0.17	18201	Ambient air pollution is not expected to give any severe symptoms	2
Cancer	Person years	0.2	21413	Mostly lung cancer, but other forms exist too	2
Skin cancer	Person years	0.05	5353	Melanoma and other skin cancers	2
Low vision	Person years	0.17	18201	Decreased stratorpheric ozone layer	2
Poisoning	Person years	0.6	64240	Largely from pesticides	1.2
Intellectual disability: mild	person years	0.031	3319	Cause by lead and mercury	4
Osteoporosis	per case	1.28	137045	Caused by Cd	3
Renal dysfunction	per case	0.64	68523	Caused by Cd and Hg	2

means full health. Weighting factors are developed for many diseases and its sequelae. Here we have only use those which are representative for health effects of environmental impacts. Table 5.10 shows the weighting factors for health impacts valued in this book.

The weighting factors for malnutrition, asthma cases, chronic obstructive pulmonary disease, low vision, and poisoning are taken from the WHO 2004 update [41]. The factors for diarrhea, malaria episodes, gravation of angina pectoris, cardiovascular diseases and infarcts are from the 2010 update for the global burden of disease [42]. The factors for osteoporosis and renal dysfunctions are from [43].

Uncertainty factors are derived from published variations in the weighting factors and estimations of how well the diseases and sequalae of published weighing factor coincides with those caused by environmental impacts.

References

1. *Producer Prices - Annual.* [cited 2019 01-04]; Available from: http://www.fao.org/faostat/en/#data/PP.
2. Tveterås S, et al., *Fish Is Food - The FAO's Fish Price Index. PLoS ONE*, 2012. **7**(5): p. 10.
3. *TIMBER database, July 2018.*, UNECE/FAO, Editor. 2018.
4. Leigh, J.H. and A.G. Randall, *Timber Trade Practice.* 4 ed. 1981, Houndsmills, Basingstoke, Hampshire and London: Macmillan Education Ltd.
5. McCarthy, D.P., et al., Financial costs of meeting global biodiversity conservation targets: Current spending and unmet needs. *Science*, 2012. **338**(6109): 946–949.
6. Hammenecker, J., *International statistics for water services*, in *IWA statistics and economics.* 2014, IWA, International Water Association.
7. Feliciano-Bruzual, C., Charcoal injection in blast furnaces (Bio-PCI): CO2 reduction potential and economic prospects. *Journal of Materials Research and Technology*, 2014. **3**(3): 233–243.
8. Lamlom, S.H. and R.A. Savidge, A reassessment of carbon content in wood: variation within and between 41 North American species. *Biomass and Bioenergy*, 2003. **25**(4): 381–388.
9. Steen, B., Calculation of monetary values of environmental impacts from emissions and resource use. the case of using the eps 2015d impact assessment method. *Journal of Sustainable Development*, 2016. **9**(6): 18.
10. *An Analysis of Energy Production Costs from Anaerobic Digestion Systems on U.S. Livestock Production Facilities.* 2007, United States Department of Agriculture, Natural Resources Conservation Service.
11. Rudnick, R.L. and S. Gao, 4.1 - Composition of the Continental Crust, in *Treatise on Geochemistry (Second Edition)*, H.D. Holland and K.K. Turekian, Editors. 2014, Elsevier: Oxford. pp. 1–51.
12. Lindberg, G., *Aluminium i Sverige.* 1973, Örebro: Ljungföretagen.
13. *5,000 Tonne per Day Open Pit Mine Model.* [cited 2019 2019-02-01]; Available from: http://costs.infomine.com/costdatacenter/miningcostmodel.aspx.
14. Jankovic, A. and W. Valery, *Reducing Grinding Energy and Cost-Magnetite Iron Ore Design Case Study*, 2010. Publisher?
15. Green, D.W. and M.Z. Southard, *Size reduction, Perry's Chemical Engineers' Handbook.* 9th edn. 2019, New York: McGraw-Hill Education.

16. Steen, B., *A systematic approach to environmental priority strategies in product development (EPS). Version 2000 – Models and data of the default method*, in *CPM reports*. 1999, Chalmers University of Technology: Gothenburg, p. 306.
17. Blatt, H., R.J. Tracy and B.E. Owens, *Petrology*. 3rd edn. 2006, New York: Freeman and Company.
18. Steen, B. and G. Borg, An estimation of the cost of sustainable production of metal concentrates from the earth's crust. *Ecological Economics*, 2002. **42**: 14.
19. Tran, T. and V.T. Luong, *Lithium Production Processes*, in *Lithium Process Chemistry*. 2015, Elsevier, p. 44.
20. Hoshino, T., *Innovative lithium recovery technique from seawater by using world-first dialysis with a lithium ionic superconductor. Desalination*, 2015. **359**: 59–63.
21. Shahmansouri, A., et al., Feasibility of extracting valuable minerals from desalination concentrate: a comprehensive literature review. *Journal of Cleaner Production*, 2015. **100**: 4–16.
22. Riley, J.P. and M. Tongudai, *The lithium content of sea water.* Deep Sea Research and Oceanographic Abstracts, 1964. **11**(4): 563–568.
23. Naidu, G., et al., Rubidium extraction from seawater brine by an integrated membrane distillation-selective sorption system. *Water Research*, 2017. **123**: p. 11.
24. Choubey, P.K., et al., Advance review on the exploitation of the prominent energy-storage element Lithium. Part II: From sea water and spent lithium ion batteries (LIBs). *Minerals Engineering*, 2017. **110**: 104–121.
25. Kushnir, D. and B.A. Sandén, *The time dimension and lithium resource constraints for electric* vehicles. *Resources Policy*, 2012. **37**(1): 93–103.
26. *Minerals database, Beryllium*. 2019 [cited 2019 2019-02-12]; Available from: https://mineralseducationcoalition.org/minerals-database/beryllium/.
27. Parthasarathy, P. and K.S. Narayanan, Hydrogen production from steam gasification of biomass: Influence of process parameters on hydrogen yield – A review. *Renewable Energy*, 2014. **66**: 570–579.
28. Ogden, J., *Prospects for Hydrogen in the Future Energy System*. 2018, Institute of Transportation Studies, University of California, Davis.
29. *Handbook of Chemistry and Physics*. 99th edn. 2018: CRC Press.
30. Popat, K.M., P.S. Anand and B.D. Dasare, *Synthesis and characterisation of boron-selective porous condensate cation exchanger. Reactive Polymers, Ion Exchangers, Sorbents*, 1988. **8**(2): 143–151.
31. Zapata, F. and R.N. Roy, *Use of Phosphate Rocks for Sustainable Agriculture*, in *Fertilizer And Plant Nutrition Bulletin*. 2004, Food And Agriculture Organization Of The United Nations: Rome.
32. Dorozhkin, S.V., Dissolution mechanism of calcium apatites in acids: A review of literature. *World J Methodol*, 2012. **2**(1): 18.
33. Economou-Eliopoulos, M. and D.G. Eliopoulos, Selenium content of sulfide ores related to *ophiolites of Greece. Journal of Environmental Pathology, Toxicology and Oncology*, 1998. **17**(3-4): 199–204.
34. Nicolaidou, A.E., Selenium content in sulfide ores from the Chalkidiki Peninsula, Greece. *Journal of Environmental Pathology, Toxicology and Oncology*, 1998. **17**(3-4): 189–197.
35. Queffurus, M. and S.J. Barnes, A review of sulfur to selenium ratios in magmatic nickel-copper and platinum-group element deposits. Ore Geology Reviews, 2015. **69**: 301–324.
36. YanardaǦ, R. and H. Orak, Total selenium concentration in various waters of Turkey. *Environmental Technology* (United Kingdom), 2001. **22**(2): 237–246.
37. Zhao, B., et al., Distribution and Accumulation of Selenium in Plants and Health Risk Assessment from a Selenium-Rich Area in China. *Polish Journal of Environmental Studies*, 2018. **27**(6): 2873–2882.

38. Torgersen, T., Continental degassing flux of 4He and its variability. *Geochemistry, Geophysics, Geosystems*, 2010. **11**(6).
39. Mohr, S. and J. Ward, *Helium Production and Possible Projection*, 2014. 4: 130–144.
40. OECD. *Labour compensation per hour worked (indicator)*. 2019 [cited 2019 15 February].
41. Who, *Global Burden Of Disease 2004 Update: Disability Weights For Diseases And Conditions*. 2004, WHO: Geneva.
42. Salomon, J.A.e.a., Common values in assessing health outcomes from disease and injury: disability weights measurement study for the Global Burden of Disease Study 2010. *Lancet*, 2012. **380**: 15.
43. Searl, A., *Exposure response functions for HM impacts on human health. Report for WP 6, EU 6th framwork project ESPREME, Estimation of willingness-to-pay to reduce risks of exposure to heavy metals and cost-benefit analysis for reducing heavy metals occurrence in Europe*. 2007.

6

Emissions of Inorganic Gases to Air

6.1 Carbon Dioxide

Carbon dioxide is a central life supporting gas. It is part of the carbon cycle, which includes production of CO_2 from respiration and consumption by photosynthesis. Changes in CO_2 concentrations and flows influence life in several ways, such as photosynthesis, respiration, climate and the state of carbonate materials.

6.1.1 Specification of the emission

Most carbon dioxide emissions are caused by combustion processes and biological respiration. Some CO_2 are also emitted from calcination of carbonates. The residence time of CO_2 in air is very long. Therefore, the exact place and time where the emission occur on the globe is not important for the effects of CO_2. Nor is the source strength for the impact per unit mass of CO_2.

The emission, whose impacts are valued here is CO_2 emissions anywhere in the world the year 2015 at any source strength.

6.1.2 Impacts pathways

As the residence time for CO_2 in the atmosphere is very long, several hundred years at least, impacts from an emission of CO_2 today should ideally be modeled for that long period. There is however no information to use such models. In the IPCC reports scenarios up to the year 2100 can be used. The impact models presented here will estimate the impacts between 2015 and 2100 caused by 1 kg of CO_2 emitted in 2015 in the IPCC reference scenario RCP6.0. There are four reference scenarios in IPCC AR5 reports. RCP6.0 is the higher middle scenario with an estimated global mean temperature increase of 1.3°C to the middle century and 2.2°C to 2100.

The most significant impacts of CO_2 are caused by its contribution to the greenhouse effect. CO_2 also cause acidification of ocean waters and rain water, resulting in corrosion of carbonate structures, like coral reefs and buildings. Information about CO_2's impacts are mostly extracted from IPCCs 5th assessment reports.

Impacts for which costs are estimated are shown in Table 6.1

Death rates in cardiovascular diseases vary with temperatures. At high temperatures there are excess mortality due to heat stress. In particular high temperatures at night are problematic for persons with cardiovascular diseases. There is no rest after a hot day. At low temperatures there is much tension that worsens the sequalae of cardiovascular diseases. There seems to be a minimum risk at about 24°C.

Climate change influences crop production in various ways. The effect is different in different areas and for different crop. In some areas there is an increased production capacity due to higher temperatures and more rain, in other areas increasing draught decrease harvests. Those who are poorest suffer most from decreased harvests. In IPPC's latest (the 5th assessment report) the problem with starvation had decreased compared to the time when the 4th assessment report was written. Poverty had decreased. Improved economy made it possible for many to buy food on the market when the crop yield was too low.

Rising sea levels and increased frequencies of high winds and rain cause loss of land areas and increased flooding, with casualties and damage on buildings and agriculture. It also causes infiltration of bacteria in water wells, which in turn increases diseases such as diarrhea.

Heat stress is a problem also for healthy workers. Working capacity decreases with high temperature. The problem gets aggravated with high humidity.

There are several ways CO_2 affects wood production. Wood production capacity increase at boreal and temperate latitudes due to higher temperatures, precipitation and CO_2 fertilization. The production capacity decreases in tropical forest. IPCC concludes in AR5 that the net impact is zero.

Table 6.1. Environmental goods affected by carbon dioxide emissions.

Environmental good	Impact indicator	Unit	Pathway
human health	YLL	personyears	heat stress
human health	YLL	personyears	cold moderation
human health	YLL	personyears	undernutrition
human health	YLL	personyears	flooding
human health	YLL	personyears	diarrhoeal diseases
human health	undernutrition	personyears	food supply
human health	working capacity loss	personhours	heat stress
human health	diarrhea	personyears	drinking water
crop	production capacity	kg	climate change
crop	production capacity	kg	rise of sea level
meat	production capacity	kg	draught
fish	production capacity	kg	ocean acidification
wood	production capacity	m^3	climate change
drinking water	production capacity	m^3	climate change
biodiversity	share of threat to redlisted species	dimensionless	habitat change

6.1.3 Environmental impact factors

6.1.3.1 YLL via heat stress and cold moderation

IPCC AR5 refers to several studies on extreme heat and concludes that it is very likely that health effects from these will increase [1]. It has long been known that an increased temperature during heat waves causes excess mortality in urban areas. The effect becomes statistically significant at daily average values around 30°C [2]. Persons older than 65 years suffering from cardiovascular diseases are at high risk.

An increased absorbency of sun radiation and lack of shadowing and evaporative cooling from vegetation largely influence the urban climate in tropical and subtropical areas.

The increased absorbency, the large heat storing capacity in buildings, the increased burning of fossil fuel and decreased cooling capacity create a heat island in the center of a city. The larger the city and the denser the population, the more expelled is the heat island compared to the surroundings. However large differences occur between different regions. In tropical humid regions buildings are mainly made of light materials to promote ventilation, while in tropical dry areas with large temperature variations between day and night the buildings are made of heavy stone, concrete, bricks, etc., to level out indoor temperature [3].

The heat island problem may increase substantially in the future, depending on several factors. The population growth and migration to urban areas increase the number of those exposed. In turn, the growth of the urban complexes increases the excess temperature. Adding global warming on top of this will enhance the effects further.

In IPCC AR5 there is no quantitative estimation of the extent of the problem with heat stress and cold moderation.

A possibility to estimate the effects of elevated temperature is to compare death statistics with the average monthly temperatures. Weihe[2] refers to a study of Sakamoto-Moniyama who correlated monthly mortality rates from cerebrovascular diseases with mean monthly temperatures for a number of countries. The relative mortality rate seems to vary linearly with temperature and has a minimum at 24°C. The excess mortality at 26–28°C was the same as at 6–8°C. Combining this information with data given for Cairo the dose-response-curve in Figure 6.1 was constructed.

To estimate the total YLL, which an increase of global average temperature would cause, one has to estimate the average shortening of life for those individuals who die during heat periods, the number of persons exposed to high temperatures and the time per year they are exposed to various temperatures.

Statistics from the heat wave in France 2003, indicate that the average lost life expectancy was 1 month [4]. The increase mortality rate above 30 degrees would be 0.3 per degree or 0.39 for the average temperature increase for the scenario RCP6.0. Using one month per death, temperature increase of 1.3 degrees, the total YLL during 2015 to 2100 are shown in Table 6.2 for various regions and seasons.

The slope in Figure 6.1 is –0.04 relative mortality rate per degree below 24°C, and 0.3 relative mortality rate per degree above 24°C. The

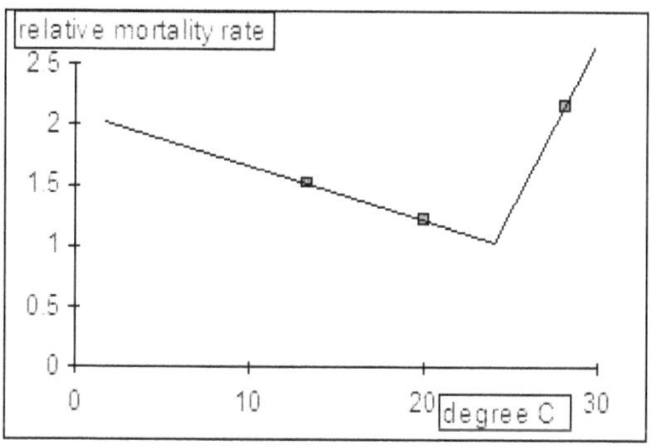

Figure 6.1. Monthly mean excess mortality as a function of temperature. The mortality at 24°C is set equal to 1.

Table 6.2. Increased and decrease mortality in different regions.

Part of world	Region	Estimated population 2025 (million)	Average monthly temperature (deg. C).				Excess mortality YLL/quarter			
			Jan–March	April–June	July–Sep	Oct–Dec	Jan–March	April–June	July–Sep	Oct–Dec
Europe	west	150	5	14	17	7	–2257	–2257	–2257	–2257
	south	159	11	15	23	14	–2392	–2392	–2392	–2392
	east	131	–10	8	15	–8	–1971	–1971	–1971	–1971
	north	84	2	9	14	6	–1264	–1264	–1264	–1264
Africa	west	558	27	26	25	25	62969	62969	62969	62969
	east	537	25	36	33	26	60599	60599	60599	60599
	north	261	14	20	25	18	–3927	–3927	29453	–3927
	central	170	26	27	26	25	19184	19184	19184	19184
	south	91	17	14	14	18	–1369	–1369	–1369	–1369
Latin America	middle	779	19	22	22	20	–11721	–11721	–11721	–11721
	tropical south	223	26	24	24	25	25165	25165	25165	25165
	temperate	429	22	15	12	17	–6455	–6455	–6455	–6455
	Carribean	58	20	20	20	20	–873	–873	–873	–873
North America		345	4	17	22	18	–5191	–5191	–5191	–5191
East Asia	China	1475	8	22	24	14	–22193	–22193	166450	–22193
	Japan	132	7	18	24	14	–1986	–1986	14896	–1986
	rest	114	0	17	21	4	–1715	–1715	–1715	–1715

South Asia	south central	1855	24	33	29	22	209332	209332	209332	–27911
	southeast	688	25	28	27	25	77639	77639	77639	77639
	southwest	271	8	20	26	10	–4078	–4078	30582	–4078
Oceania	Australia/N Z	27	21	16	14	18	–406	–406	–406	–406
	Melanesia	10	20	20	20	20	–150	–150	–150	–150
	M&Polynesia	1	20	20	20	20	–15	–15	–15	–15
Former Sovjet		368	–10	8	15	–8	–5537	–5537	–5537	–5537
		8916				SUM	381386	381386	654950	144144

average increase in temperature in the RCP6.0 is 1.3°C giving a net change in relative mortality of –0.052 and 0.39 respectively. The average mortality rate can be estimated as 1000000/(72*4) = 3472 cases per million inhabitants and quarter, where 72 years is the average life expectancy.

The sum of increased YLL is 1.85E+06 and the sum of decreased YLL is 2.90E+05 per year. The net YLL is 1.56E+06 per year. For the whole period 2015–2100 the increased mortality is 1.57E+08 YLL, the decreased mortality 2.46E+07 YLL and the net effect is 1.33E+08 YLL.

The contribution to the mortality changes is the inverse of the added emissions causing them. By the year 2100 the accumulated CO_2 emissions are estimated by IPPC to be 5920 Gton [5] and CO_2's contribution to temperature raise 88% [6]. The contribution of 1 kg of extra CO_2 is therefore 1.69E–16. The environmental impact factor for heat stress is 1.57E+08*1.69E–16 = 2.66E–08 YLL/kg CO_2 and for cold moderation –2.46E+07*1.69E–16 = –4.16E–09 YLL/kg CO_2.

Several factors contribute to the uncertainty of the environmental impact factors. The dose-response function in Figure 6.1 is probably not linear and it lacks one important parameter: humidity. The population figures in Table 6.2 is from around the year 2000. More recent estimates indicate a higher population growth in Africa and ideally the population forecast would be for 2050–2055. The differences in estimated global mean temperature from different models and scenarios [6] is in the order of a factor of 1.4 around the 1.3 degrees temperature increase used here. As the dose-response curve is assumed to be linear, the same uncertainty in response applies. A net uncertainty of a factor of 1.6 seems to be a reasonable estimate.

6.1.3.2 YLL via undernutrition

Nutrient deficiency was registered as the cause of death for 684 000 persons in 2010 [7]. UNICEF claims that three million children under 5 died from malnutrition. The reason for the mortality rate for children being much higher than for the total population may be explained by that undernourished children are susceptible to several other diseases, which may or may not be registered as death causes. Assuming 27 years reduction in life expectancy for adults and 70 for children, based on conditions in the poorest countries [8], we get a total life expectancy decrease of 0.684*27 + 3*70 = 228 million YLL/year. No quantitative estimation of how climate contributes to this increased mortality has been found in scientific literature. Lloyd et al. [9] found an increase in severe stunting among children to 23% (central sub Saharan Africa) to 62% (South Asia) to 2050. The average, 42% is used as an indicator for climate-caused increased mortality. Thus, we get 288*0.42*85 = 10282 million YLL for the period 2015–2100.

The contribution of 1 kg of extra CO_2 is the same as for heat stress and cold moderation, 1.69E–16. The environmental impact factor therefore becomes 1.03E+10*1.69E–16 = 1.47E–06 YLL/kg CO_2.

The estimation of the relative increase in mortality due to climate change is highly uncertain. The assumption that it correlates to the increase in severe stunting seems reasonable, but if the regression slope is 1, it is questionable. An alternative way of estimating the increased risk for starvation and immature death would be to compare with the decrease in crop production, 10%. As this happens in poor countries, 10% should be a lower estimate of the increase. Based on this, the uncertainty of the environmental impact factor is set to a factor of 2.

Starvation rates are highly dependent on socioeconomic factors, which are extremely difficult to predict, but that kind of uncertainty is less relevant here. As mentioned in the chapter on goal and scope, the intention is not to foresee the future, but to find out the consequences of "business as usual". If you want to contribute to sustainable development, there is no point in assuming it will happen anyway.

6.1.3.3 YLL via flooding

IPCC AR5 WG2 refers to Dasgupta et al. (2009) who "developed a spatially explicit mortality model for 84 developing countries and 577 coastal cities. They modelled 1:100 year storm-surge events, and assessed future impacts under climate change, accounting for sea level rise and a 10% increase in event intensity. In the 84 developing countries, an additional 52 million people and 30,000 km^2 of land were projected to be affected by 2100."

IPCC also reports that "It has been estimated conservatively that around 2.8 billion people were affected by floods between 1980 and 2009, with over 500,000 deaths". This means that the ratio of mortality to affected people is 0.000179. Even if this ratio is not measured for a 1:100 years storm-surge event, and not for the period up to 2100, it is the only one found and assumed to be relevant in these cases too. So, the extra 52 million affected during 2015–2100 is estimated to result in 52*0.000179*85/29 = 0.027 million deaths or 0.982 million YLL if the risk for causalities are the same for all ages, and the average reduction in life expectancy is 36 years.

The contribution of 1 kg of extra CO_2 is the same as for heat stress and cold moderation, 1.69E–16, and why the environmental impact factor becomes 982000*1.69E–16 = 1.66E–10 YLL/kg CO_2.

The uncertainty in this kind of estimations is high, at least a factor of 3, but this is unlikely to significantly contribute to the overall uncertainty in estimation of the monetary value of environmental impacts from emissions of CO_2, as the impact factor is small.

6.1.3.4 YLL via diarrheal diseases

Kolstad and Johansson [10] projected an increase in the risk of diarrhea in the tropics and subtropics by 2010–2039 of 8–11% and 22–29% by 2070–2099 due to global warming increasing the temperature with 4 degrees to the end of the century. This will correspond to the RCP8.5 scenario rather than the RCP6.0 which is used as default here. The lower estimate of 10% is therefore assumed to be valid as an average for the period 2015–2100.

During 2010, 1.4 million persons died due to diarrheal diseases [7]. The increase for 2015–2100 will therefore be 0.1*1.4*85 = 11.9 million deaths. Children are particularly vulnerable, why 60 years reduction in life expectancy is assumed as an average. In terms of YLL, we obtain 11.9*60 = 714 million YLL.

The contribution of 1 kg of extra CO_2 is the same as for heat stress and cold moderation, 1.69E–16, why the environmental impact factor becomes 7.14E+08*1.69E–16 = 1.21E–07.

Socioeconomic development and shift of scenario from RCP8.5 to RCP6.0 are mostly contributing to the uncertainty, which is set to a factor of 3.

6.1.3.5 Undernutrition via decreasing food supply

Climate change will cause an increase in the number of undernourished children under 5 to 25 million by the year 2050, according to models referenced in WGII AR5, Chapter 11, Table 11.2. This is an increase of 22%. These models do not consider socioeconomic development. Studies including socioeconomic scenarios have indicated a 10% increase. Children under 5 is about 10% of the population in developing countries. It is assumed that the rest of the family also is starving, i.e., total extent of malnutrition due to climate change is 120 million per year or 10200 million persons for the period 2015–2100.

The contribution of 1 kg of extra CO_2 is the same as for heat stress and cold moderation, 1.69E–16, why the environmental impact factor becomes 1.02E+10*1.69E–16 = 1.72E–06 person-years/kg CO_2 for the period 2015–2100.

The uncertainty in the environmental impact factor depends largely on unknown socioeconomic development and on choice of reference scenario. As mentioned above, adding a socioeconomic scenario reduced the estimate with a factor of 2. Besides, the reference scenario used corresponded to RCP8.5 rather than RCP6.0 which is used as a default here. An uncertainty factor of 2 is therefore assumed. That would mean that the true value would with 95% probability lie within two standard deviations in the log-normal distribution, i.e., a factor of 4.

6.1.3.6 Decreased working capacity from heat stress

AR5 WGII, Chapter 11.6.2.4 cites an estimation from Dunne et al. [11] of a 20% loss of productivity globally in RCP4.5 by 2100 compared to 1990 and 30% for RCP8.5. From Dunne et al. (their Figure 2), the decrease in productivity for the global labor force (performing physical labor) is estimated to be 6% as an average for 2015–2100 for RCP6.0. Dunne et al. estimate the global population weighted reduction in working capacity to between 2 and 20%, so 6% is an average in a log-normal distribution with an uncertainty factor of 1.8. The total workforce with a global population of 9 billion is estimated to be 6*0.65*0.3 = 1.17 billion, where 6 is the population in billions in ages 20–65 years, 0.65 is the approximate present employment rate in OECD countries and 0.3 is the share of the work force performing physical labor (estimates from ILO database ILOSTAT and the OECD labor market outcomes—employment rates). However, for individuals also non-paid labor has a value. Assuming the same share of physical labor, the affected population will be 1.8 billion. The model by Dunne et al. does not cover non-paid labor outside working hours. One would expect that the reduction of labor capacity is less severe in the mornings and evenings but the same during weekends.

The total reduction in working capacity for the years 2015–2100 is set to 1.8E+09*0.06*365*8*85 = 2.68E+13 person-hours. The contribution of 1 kg of extra CO_2 is the same as for heat stress and cold moderation, 1.69E–16, and the environmental impact factor becomes 2.68E+13*1.69E–16 = 4.53E–03 person-hours/kg CO_2.

Another estimation of the loss of productivity is made by Kjellstrom et al. [12]. They found a reduction of as much as 11–27% in the A2 scenario in South-East Asia, Andean and Central America, and the Caribbean. The A2 scenario used in IPCC assessment reports AR3 and AR4, is regionally oriented with increased warming of 2.0–5.4°C. Kjellstrom et al. also analyzed the sensitivity to changes in work conditions. Then the average reduction in work capacity changed from 6 to 2% when changing work conditions were included in their modelling. An uncertainty of a factor of 2 is therefore assumed.

6.1.3.7 Diarrhea via climate change and degraded water quality

"In countries with endemic cholera, there appears to be a robust relationship between temperature and the disease" (AR5, WGII, Ch 11.5.2). Kolstad and Johansson [10] projected an increase of 8–11% for the risk of diarrhea in the tropics and subtropics in 2039 and 22–29% by 2070–2099 due to climate change, using the A1B scenario (1.4–6.4°C). The study did not account for future changes in economic growth and social development

(AR5 WGII Ch 11.5.2.3). According to Vos et al. [13], the extent of diarrheal diseases in 2010 was 8 million personyears, probably mostly located in the regions studied by Kolstad and Johansson.

If the lower values are used, an average of 15% increase for the period 2015–2100 is obtained. Adjusted for population growth the increase in diarrheal diseases will be 8E+06*9.7/7.3*0.15 = 1.59E+06 person-years. The contribution of 1 kg of extra CO_2 is the same as for heat stress and cold moderation, 1.69E–16. The environmental impact factor will therefore be 1.59E+06*1.69E–16 = 2.69E–10 person-years/kg CO_2.

The same uncertainty is assumed as for YLL from diarrheal diseases, i.e., a factor of 3.

6.1.3.8 Decreased crop production capacity via increased draughts

FAO [14] reports a global crop production of 1.06E+10 tons for 2015. IPCC WGII AR5 Chapter 7, Figure 7.7 indicates an average 5% decline (1% per decade) until 2100 for major cereals due to increased droughts. This decline estimate is assumed to be relevant for other crops too. The world production is assumed to follow population growth. This means that the loss of crop production capacity is 1.06E+10*9.7/7.3*85*0.05 = 5.99E+10 ton or 5.99E+13 kg between 2015 and 2100. The contribution of 1 kg of extra CO_2 is the same as for heat stress and cold moderation, 1.69E–16. The environmental impact factor becomes 5.99E+13*1.69E–16 = 1.01E–02 kg crop/kg CO_2.

Unknown degrees of adaptation and climate scenario adds to the uncertainty, as well as the sensitivity of different crops to climate change. A factor of 3 is assumed.

6.1.3.9 Decreased crop production capacity due to rising sea level

Bosello et al. estimates land loss from sea level rise to 1.25E+05 km^2. According to FAO [14] the crop yield for different crops varies between half a ton and 48 tons per ha (for potatoes). The average yield for rice, which is 3.9 ton/year will be used. The temporal extent of the impact is set to 500 years. The added impact is then 1.25E+05*100*3900*500 = 2.44E+13 kg crop. The contribution of 1 kg of extra CO_2 is the same as for heat stress and cold moderation, 1.69E–16. The environmental impact factor becomes 2.44E+13*1.69E–16 = 4.12E–03 kg crop/kg CO_2.

The IPPC AR5 WG1 table SPM2, presents an uncertainty range for the sea level rise between 0.18 and 0.32 m for the years 2046–2065 and 0.33–0.63 m for the years 2081–2100 for the RCP6.0 scenario. Bosello et al.

assumes a sea level rise of 1m, which means that their area estimations are too high for the 2100 century, but the best available for a 500 years perspective. Given the long duration of impacts and lack of impact models with 500 years' perspective, an uncertainty of a factor of 3 is assumed.

6.1.3.10 Decreased meat production capacity from draught

FAO statistics [14] show a current total production of 0.39 billion tonnes of meat. Assuming IPCC WGII AR5 Chapter 7.4.3 estimation of an average 5% decline of crop being representative also for meat (1% per decade), the decreased meat production capacity becomes 0.39*9.7/7.3*0.05*85 = 2.20 billion ton for the period 2015–2100. The world production is assumed to follow population growth. Hence the correction factor 9.7/7.3.

The contribution of 1 kg of extra CO_2 is the same as for heat stress and cold moderation, 1.69E–16. The environmental impact factor becomes 2.20E+12*1.69E–16 = 3.72E–04 kg meat/kg CO_2.

Uncertainty is the same as for crop production capacity, a factor 3.

6.1.3.11 Decreased fish production capacity from ocean acidification

About half of the worlds fish catch comes from fishing and half from aquaculture. No comprehensive and quantitative estimation of the impact of climate change on fish production capacity has been found. IPCC AR5 WG2 reports about both positive effects and negative. The only quantitative estimation found is on mollusk production, which declines due to ocean acidification.

The total harvest of mollusks was 13.6 million tons/year as an average 1997–2006 [15]. A decrease in calcination of 28% and survival of 30% is estimated for the IPCC IS92a business-as-usual scenario for the year 2100 by Kroeker et al. [16]. With unchanged production and an average decline of 15% this will indicate a loss of 13.6*0.15*85 = 173 million tons for the period 2015–2100.

The contribution of 1 kg of extra CO_2 is the same as for heat stress and cold moderation, 1.69E–16. The environmental impact factor becomes 1.73E+11*1.69E–16 = 2.93E–05 kg fish/kg CO_2.

The effect of ocean acidification varies for different species. Kroeker et al. find a variation between 5 and 60% on survival. As the mollusk harvest is made up by many species, the contribution to the uncertainty in the environmental impact factor is less, but considering that the IS92a scenario results in a higher temperature than our RCP6.0, an uncertainty factor of 2 seems reasonable.

6.1.3.12 Decreased wood production capacity via climate change

Forests cover about 4 billion hectares of land. An average production capacity is estimated to be 6 m^3 wood/hectare and the average impact from climate change and increase of plus or minus a few % globally. Kramer et al. [17] indicate a growth of about 5% for boreal forests, while temperate and tropical forests may even be negatively influenced. AR5 WGII, Chapter 4, reports on very different results for different areas and time periods in all three forest types (boreal, temperate and tropical). A 2% change would give a change of 0.48 billion m^3/year in production. We assume an average of 0 and a linear normal distribution for the uncertainty with a standard deviation of 2%.

The contribution of 1 kg of extra CO_2 is the same as for heat stress and cold moderation, 1.69E–16. The environmental impact factor is 0 ± 0.48E+09*85*1.69E–16 = 0 ± 6.90E–06 m^3 wood/kg CO_2.

6.1.3.13 Decreased drinking water production capacity via climate change

AR5 WGII, Chapter 3.4.4 says: "Each degree of global warming (up to 2.7°C above pre-industrial levels) is projected to decrease renewable water resources by at least 20% for an additional 7% of the world population"[18]. For RCP 6.0 this means about 1°C as an average from 2015 to 2100, and that 0.07*9 billion persons living with water scarcity is affected. The average water withdrawal today is around 400–700 m^3/person and year in countries at risk, like Mexico, India and China. The decreased availability of water is therefore estimated to be 0.07*9E+09*500 *0.2 = 6.3E+10 m^3/year and 5.36E+12 m^3 for the period 2015–2100. It is unclear what part of this is drinking water and why 500 m^3 is used as an upper limit.

The contribution of 1 kg of extra CO_2 is the same as for heat stress and cold moderation, 1.69E–16. The environmental impact factor is therefore 5.36E+12*1.69E–16 = 9.06E–04 m^3/kg CO_2 at most.

The uncertainty is high, but as the 9.06E–04 m^3/kg CO_2 is an upper limit, the uncertainty is only towards lower values.

6.1.3.14 Decreased biodiversity via climate induced habitat change

70% of the present population of birds will be affected due to changing habitats (AR5 WGII).

Future conservation measures need to be increased and climate change alone is assumed to require doubling of means. There are no comprehensive estimations found of the global climate change impacts on biodiversity, but only on local habitats. As a best estimate the impact on bird population will be used for assuming a doubling of the impact between 2015 and 2100 in terms of resources needed to preserve biodiversity. This means that the environmental impact factor is 1.69E–16 shares of the present impact on biodiversity per kg of CO_2. The uncertainty is assumed to be a factor of 4.

6.1.4 Summary and monetary valuation of CO_2 emissions

A summary of environmental impact factors and monetary impact values are shown in Table 6.3. When comparing different monetary impact values, we find that YLL from undernutrition and personyears loss of working capacity are contributing to 91% of the total value. YLL from diarrheal diseases and undernutrition from decreasing food supply account for another 6%. The contribution from flooding is surprisingly low but may partly be explained by the fact that damage to buildings and infrastructure is not included. The focus on long-term and satisfiers for basic needs disregards many instrumental resources and resources not part of the natural capital.

6.2 Carbon Monoxide

6.2.1 Specification of the emission

The major sources of atmospheric CO are oxidation of hydrocarbons (mostly CH_4 and isoprene) and direct emission resulting from incomplete combustion of biomass and fossil fuels [6]. The residence time of CO in air is one to three months [6, 19]. Direct emissions from traffic is of special interest as they cause local health effects. The emission whose impacts are valued here is CO emissions anywhere in the world the year 2015 at any source strength.

6.2.2 Impact pathways

CO does not give a direct impact on radiative forcing and climate but participates in atmospheric reactions influencing formation of oxidants and the residence time for methane and NMVOC. CO is toxic to humans and causes direct health hazards. The impact models presented here will estimate the impacts caused by 1 kg of extra CO emitted in 2015 in the IPCC reference scenario RCP6.0. RCP6.0 is the higher middle scenario with an estimated global mean temperature increase of 1.3°C to the middle of the 21st century and 2.2°C to 2100. Impacts for which costs are estimated are shown in Table 6.4.

Table 6.3. Summary of environmental impact factors and monetary impact values of CO_2 emissions.

Environmental good	Impact indicator	Unit	Pathway	Environmental impact factor	Uncertainty	Indicator value, ($/unit)	Impact value ($/kg CO_2)
human health	YLL	personyears	heat stress	2.65E–08	1.6	107,067	2.84E–03
human health	YLL	personyears	cold moderation	–4.16E–09	1.6	107,067	–4.45E–04
human health	YLL	personyears	undernutrition	1.74E–06	2	107,067	1.86E–01
human health	YLL	personyears	flooding	1.66E–10	3	107,067	1.78E–05
human health	YLL	personyears	diarrhoeal diseases	1.21E–07	3	107,067	1.29E–02
human health	undernutrition	personyears	food supply	1.72E–06	4	6,424	1.11E–02
human health	working capacity loss	personhours	heat stress	4.53E–03	2	30	1.36E–01
human health	diarrhea	personyears	drinking water	2.69E-10	3	11,242	3.02E–06
crop	production capacity	kg	climate change	1.01E–02	3	0.289	2.93E–03
crop	production capacity	kg	rise of sea level	4.12E–03	3	0.289	1.19E–03
meat	production capacity	kg	draught	3.72E–04	3	2.59	9.63E–04
fish	production capacity	kg	ocean acidification	2.92E–05	2	2.42	7.08E–05
wood	production capacity	m^3	climate change	0.00E+00	6.90E–06	72	0.00E+00
drinking water	production capacity	m^3	climate change	9.06E–04	3	1.87	1.69E–03
biodiversity	share of threat to redlisted species	dimensionless	habitat change	1.69E–16	4	7.61E+10	1.29E–05
						Sum	3.56E–01

Table 6.4. Pathways by which carbon monoxide, CO, impact on environmental goods.

Environmental good	Impact indicator	Unit	Pathway
human health	YLL	personyears	climate change
human health	YLL	personyears	oxidant formation
human health	gravation of angina pectoris	personyears	direct exposure
human health	undernutrition	personyears	climate change
human health	working capacity loss	personhours	climate change
human health	diarrhea	personyears	climate change
crop	production capacity	kg	climate change
crop	production capacity	kg	oxidant formation
meat	production capacity	kg	climate change
fish	production capacity	kg	climate change
wood	production capacity	m^3	climate change
drinking water	production capacity	m^3	climate change
biodiversity	share of threat to redlisted species	dimensionless	climate change

Impacts due to climate change are the same kinds as those caused by CO_2.

CO contributes to formation of oxidants, which are causing acute excess mortality and decreased crop harvests. CO is toxic to humans as it blocks the transport of oxygen with blood hemoglobin and increase attacks and symptoms for persons with angina pectoris.

6.2.3 Environmental impact factors

6.2.3.1 YLL via climate change

CO has a global warming potential of between 1.6 and 7.6 depending on where it is emitted in the world and how much of aerosol formation mechanisms that are included in the modelling [6]. Here, we use a GWP of 3.5 as a best estimate with an uncertainty of a factor of 1.5. This means that the extent of the various impacts is the same as for CO_2, but the contributions from 1 extra kg of CO is 3.5 times higher, i.e., 5.92E–16 per kg.

The environmental impact factor for all CO_2 pathways to YLL becomes (1.57E+08 – 2.46E+07 + 1.03E+10 + 9.82E+05 + 7.14E+08) *5.92E–16 = 6.59E–06 YLL/kg CO with an uncertainty factor of 1.6.

6.2.3.2 YLL via exposure to ozone

Lelieveld et al. [20] calculated global, regional and megacity premature mortality due to particles and ozone. They found that 5.215 E+06 YLL was caused by ozone in 2005.

The contribution from CO to ground level ozone formation may be estimated by multiplying the emissions of all ozone-forming substances (AR5, WGI, Table 8 SM.19), mainly NO_x (1.22E+05 Gg/yr), CO (8.93E+05 Gg/yr), NMVOC (1.6E+05 Gg/yr) and methane (3.64E+05 Gg/yr) with their respective POCP.

The global emission of CO from is estimated to be 974 Tg [21] for 1990. Most CO is emitted from biomass burning. Remote sensing studies of emissions from biomass burning gave results between 365 and 1422 Tg CO per year [22]. A best estimate of 1E+12 kg will be used for the modelling.

CO's contribution to global ground level ozone is estimated to as 6.36% (Table 6.5).

The differences seen in Table 6.5 between the POCP-based calculations and the calculations reviewed by Stevenson et al. seem reasonable. For NO_x, where the residence time is relatively short, much of the ozone generation takes place in the trajectories for which POCP are modeled. 16.3% of the sum of anthropogenic and naturally generated ozone is about the same as 31% of the anthropogenic generated ozone. Carbon monoxide also has a relative short residence time, about 0.1 years [19]. 6.4% contribution to total ozone generation is about the same as 15% contribution to anthropogenic ozone. For methane, the picture is quite different. Methane has a long residence time in air and can be expected

Table 6.5. Global average contributions by different photo-oxidant creating substances to ground level ozone concentrations. Photo oxidant creation potentials (POCP) are taken from Labouze et al. [23] and is an average of their $POCP_{AOT60}$ and $POCP_{AOT90}$, which only considers ozone concentrations above 60 and 90 ppb and are relevant for health effects. Global emissions of NO_x are found in IPCC AR5, supporting material, Table 8 SM.19. Global emissions of CH_4 is given by Tsuruta et al. [24], emissions of NMVOC by Crippa et al. [25] and of isoprene and monoterpenes by IPCC AR5 [6]. Data from Stevenson et al. [26] are given for comparison. They estimated anthropogenic contributions to radiative forcing by stratospheric ozone.

Substance	POCP	Global emission (kg)	O3 creation	Contribution	Stevenson et al.
NOx	0.46	1.22E+11	5.62E+10	16.3%	31%
CO	0.022	1.00E+12	2.20E+10	6.4%	15%
CH4	0.008	5.16E+11	4.13E+09	1.2%	44%
NMVOC	0.33	1.70E+11	5.61E+10	16.2%	9%
Isoprene	0.34	5.72E+11	1.95E+11	56.3%	
monterpenes	0.15	8.50E+10	1.27E+10	3.7%	
		sum	3.46E+11		

to be well mixed in the atmosphere before generating ozone. The four or five days modeled for calculating POCPs is not representative for the ozone contributing to radiative forcing. So, methane has a much higher contribution to the ozone column in the atmosphere than to ground level ozone in populated areas. Only a part of the NMVOC is of anthropogenic origin and this may explain why Stevenson et al. find a much lower value for NMVOC contribution to ozone generation.

The total contribution from 1 kg of extra CO emitted is thus 1/1.00E+12*0.0636 = 6.36E–14, and the environmental impact factor is 5.215E+06*6.36E–14 = 3.32E–07 YLL/kg CO.

Using European POCP data for global conditions introduces significant uncertainties. Cheng et al. [27] determined POCP for isoprene and α-pinene to 1.71 and 1.52 respectively for Chinese conditions. This would reduce the CO contribution to less than 2%. This indicates an uncertainty of a factor of 2.

6.2.3.3 Decreased crop productivity caused by ozone

Van Dingenen et al. [28] estimate the global loss of agricultural crop for the year 2000 due to ozone to be between 45 and 82 million metric tons of wheat, and 17–23 million metric tons for rice, maize and soybean. Avnery et al. estimated the global crop loss to be between 79 and 121 million metric tonnes. A best estimate of 90 million tonnes is used and an uncertainty factor of 1.5.

The contribution to crop damaging ground level ozone can be calculated in the same way as in Table 6.5, but with POCP measures for accumulated ozone above 40 ppb, $POCP_{AOT40}$ [23], which are slightly different the contribution becomes 5.87E–14, and the environmental impact factor 9.0E+10*5.87E–14 = 5.28E–03 kg crop/kg CO.

The uncertainty is thought to be the same as for the environmental impact factor for YLL from ozone, i.e., a factor of 2.

6.2.4 Summary and monetary valuation of CO emissions

A summary of environmental impact factors and monetary values of environmental impacts is shown in Table 6.6.

Climate change impacts contributes to 94% to the environmental damage cost of CO, and as for CO_2 it is undernutrition and decreased working capacity that is the major causes. Ozone creation and aggravation of angina pectoris contributes each with 3%.

Table 6.6. Summary of environmental impact factors and monetary impact values of 1 kg of CO emission.

Environmental good	Impact indicator	Unit	Pathway	Environmental impact factor	Uncertainty	Indicator value ($/unit)	Impact value ($/kg CO)
human health	YLL	personyears	climate change	6.59E–06	1.6	107,067	7.06E–01
human health	YLL	personyears	oxidant formation	3.32E–07	1.6	107,067	3.55E–02
human health	gravation of angina pectoris	personyears	direct exposure	4.22E–07	3	107,067	4.52E–02
human health	undernutrition	personyears	climate change	6.03E–06	4	6,424	3.88E–02
human health	working capacity loss	personhours	climate change	1.59E–02	2	30	4.76E–01
human health	diarrhea	personyears	climate change	9.40E–10	3	11,242	1.06E–05
crop	production capacity	kg	climate change	5.39E–02	3	0.289	1.56E–02
crop	production capacity	kg	oxidant formation	5.28E–03	3	0.289	1.53E–03
meat	production capacity	kg	climate change	1.30E–03	3	2.59	3.37E–03
fish	production capacity	kg	climate change	1.02E–04	2	2.42	2.48E–04
wood	production capacity	m^3	climate change	0.00E+00	3.45E–03	72	0.00E+00
drinking water	production capacity	m^3	climate change	3.17E–03	3	1.87	5.93E–03
biodiversity	share of threat to redlisted species	dimensionless	climate change	5.92E–16	4	7.61E+10	4.50E–05
						Sum	1.33E+00

6.3 Nitrogen oxides, NO_x

Nitrogen oxides, NO_x, is the sum of NO and NO_2. Like CO_2, they are part of a life supporting cycle, the nitrogen cycle. The reason for considering them together is that they transform into each other. The most common reaction is NO oxidation to NO_2 in air. N_2O, nitrous oxide, is also a nitrogen oxide but is assessed separately. The fate of anthropogenic nitrogen has been reviewed by Schlesinger [29].

6.3.1 Specification of the emission

NO and NO_2 are formed naturally in lightnings and by biological processes. Anthropogenic sources are mainly combustion processes. Most of the NO_x formed in a combustion process is NO. There are two mechanisms for the formation of NO_x during combustion. One is the reaction between oxygen and nitrogen in the combustion air. The other is oxidation of N in the fuel.

The flows assessed here is emissions to air, any place in the world in 2015 and at any source strength.

Any mass measures of NO_x in this book is referring to NO_2. The unit is therefore kg of NO_x as NO_2. Some NO_x-data in the IPCC AR5 reports are given per kg of N.

The emission, which impact costs are valued, is one kg extra of NO_x measured as NO_2 emitted to air 2015 anywhere in the world and at any source strength.

6.3.2 Impacts caused on environmental goods and impact pathways

Nitrogen oxides enter the atmosphere primarily as NO. They are oxidized first to NO_2 and then to organic nitrates, including peroxy nitrates and more strongly bound mono and multifunctional alkyl nitrates, and nitric acid [30]. Some NO_2 is reduced back to NO when participating in photochemical reactions. The residence time for NO_x in the atmosphere is in the order of a few days [31]. The reaction between NO and ozone is fast and equilibrium is established within hours [32].

NO_x has several types of impacts. It impacts directly and indirectly on climate, human health, acidification, eutrophication and soil fertility. Impacts on human health and eutrophication are negative and impacts on climate and fertility positive. Table 6.7 shows the pathways of NO_x's impacts on environmental goods.

NO_x has a fertilizing effect on vegetation, algae and phytoplankton but only when N is a growth limiting factor. In agriculture, N is already added in sufficient amounts and extra N from NO_x has no fertilizing effect. Some

Table 6.7. Pathways by which nitrogen oxides, NO_x, impact on environmental goods.

Environmental good	Impact indicator	Unit	Pathway
human health	YLL	personyears	secondary particles
human health	YLL	personyears	climate change
human health	YLL	personyears	oxidant formation
human health	YLD	personyears	secondary particles
human health	undernutrition	personyears	climate change
human health	working capacity	personhours	climate change
human health	diarrhea	personyears	climate change
crop	production capacity	kg	oxidant formation
crop	production capacity	kg	climate change
meat	production capacity	kg	climate change
fish	production capacity	kg	eutrophication, dead zones
fish	production capacity	kg	N-nutrification of ocean
fish	production capacity	kg	acidification
wood	production capacity	m^3	oxidant formation
wood	production capacity	m^3	N-fertilization
wood	production capacity	m^3	climate change
drinking water	production capacity	m^3	climate change
biodiversity	share of threat to redlisted species	dimensionless	climate change
biodiversity	share of threat to redlisted species	dimensionless	eutrofication
biodiversity	share of threat to redlisted species	dimensionless	acidification

of the nitrogen from NO_x is re-emitted as N_2O after deposition on soil, but this is only around 1% [33], and not included in the impact models here.

6.3.3 Environmental impact factors

6.3.3.1 YLL via exposure to secondary particles

Ever since the discovery of the chronic health effects of fine particles 1993 by Dockery et al. [34], nitrate and sulphate particles have been suspected to contribute to excess mortality as part of PM2.5 and PM10. However, in IPCC AR5 it is indicated that this might not be so. In a recent review by

Yang et al. [35] they conclude that "Our analysis suggests that black carbon and organic carbon are important detrimental components of PM2.5, while other constituents are probably hazardous to human health". In deciding whether nitrates cause increased mortality, Hill's criteria for causation [36] may be applied. They consider (1) strength, (2) consistency, (3) specificity, (4) temporality, (5) biological gradient, (6) plausibility, (7) coherence, (8) experimental evidence and (9) analogical evidence. The cases for nitrates and sulphates are summarized in Table 6.8.

Lelieveld et al. [20] calculated the world excess mortality from PM2.5 to 1.59E+07 YLL for the year 2005.

The mass fraction of nitrates in PM2.5 in ambient air is only around 6% (IPCC AR5 WG1, Chapter 7) and the global emission of NO_x is 1.22E+11 kg/year (IPCC AR5 WG1 Table 8, SM.19). Assuming a much lower health effects for nitrates than the average PM2.5 (25%), the contribution of an emission of 1 kg NO_x to the 1.59E+07 YLL/year is 0.06*0.25/(1.22E+11) = 1.23E–13 and the environmental impact factor 1.49E+7*1.23E–13 = 1.96E–06 YLL/kg NO_x.

Table 6.8. Conformity with criteria for causation of excess mortality from exposure of nitrates and sulphates.

Criteria for causation	Evaluation	Conformity
strength	Correlation is weak as is slope of regression	low
consistency	Several studies have had similar results	moderate
specificity	The impact is specific to exposed populations	moderate
temporality	The effect comes after exposure	high
biological gradient	The effect increases with increasing exposure	high
plausibility	The main toxicity mechanism is thought to be oxidative stress. PM2.5 has the ability to cross the alveoli of the lung and can finally enter into the blood stream to produce inflammatory or oxidative damage, leading to more drastic and long-term secondary effects on both cardiovascular and nervous systems[37]. There are indications in mice tests that the lung recovers after exposure ceases [38]. This would mean that the residence time of particles deposited in the lung is important and that nitrates and sulphates, which are easily dissolved are likely to have low impacts	low
coherence	There are no contradictions to current knowledge	High
experimental evidence	Inflammatory effects have been seen	Moderate
analogical evidence	Different types of particles give similar effects	High

There is an uncertainty in the dose-response function. A factor of 2 would imply that the effect is between 6 and 100% with 95% probability. This seems reasonable.

6.3.3.2 YLL via climate change

NO_x affects climate in several ways. It contributes to global warming by creating ozone. It counteracts global warming by creating nitrate particles which has a cooling effect and by decreasing the methane concentration. According to IPCC AR5 WG1, NO_x has a negative global warming potential, but "due to high reactivity and the many nonlinear chemical interactions operating on different time scales, as well as heterogeneous emission patterns, calculation of net climate effects of NO_x is difficult" (IPCC AR5 WG1 8.7). Several GWP values are presented in IPCC AR5 WG1, Table 8.A.3. ranging from –11 to –159 for GWP 100. The value –159 is the one which includes both indirect and direct aerosol processes, while the value –11 does not include aerosol processes. The problem with choosing a GWP value is a trade-off between the guiding principles of accuracy and completeness. Most GWP models do not include particles, and there is only one publication including particles. A compromise is made, the GWP –95 ± 31 will be used. It represents a model where direct effects of secondary particles are included.

The sum of the impact factors for all mortality effects of CO_2 is 1.88E–06YLL/kg CO_2. For NO_x it will then be –95*1.88E–06 = –1.79E–04 YLL/kg NO_x. The uncertainty will be high, a factor of 2 is assumed.

6.3.3.3 YLL via exposure to ozone

Lelieveld et al. [20] calculated global, regional and megacity premature mortalitydue to particles and ozone. They found that 5.215E+06 YLL was caused by ozone in 2005. Cohen et al. found that 4.1 million DALYs was caused by ozone and excess mortality from chronic obstructive pulmonary disease (COPD) in 2015. The global emission of NO_x is 1.22E+11 kg/year (IPCC AR5 WG1, Table 8, SM.19).

In Table 6.5, NO_x was estimated to contribute to 16.3% of ground level ozone.

The environmental impact factor therefore becomes 5.215E+06*0.163/ 1.22E+11 = 6.97E–06 YLL/kg NO_x.

POCP for NO_x may vary and even be negative under some circumstances [39]. An uncertainty factor of 2 is assumed for the environmental impact factor.

6.3.3.4 Respiratory and cardiovascular diseases caused by secondary particles

Cohen et al. [40] found that long term exposure to PM2.5 caused 3.9 million (2.6 million to 5.2 million) years lived with disability (YLD), in 2015.

The mass fraction of nitrates is only around 6% (IPCC AR5 WG1, Chapter 7) and the global emission of NO_x is 1.22E+11 kg/year (IPCC AR5 WG1 Table 8, SM.19). Assuming a much lower health effects for nitrates than the average PM2.5 (25%), the contribution to the 1.59E+07 YLL/year is 0.06*0.25/(1.22E+11) = 1.23E–13 and the environmental impact factor 3.9E+06*1.23E–13 = 4.80E–7 YLD/kg NO_x.

The uncertainty given by Cohen et al. concerns PM2.5. There is also an uncertainty in the estimation of the toxicity of sulphates and in the contribution from NO_x-emissions. A factor of 2 is assumed for the standard deviation of the environmental impact factor.

6.3.3.5 Undernutrition via climate change

For CO_2, the environmental impact factor was determined in 6.1.3.5 to 1.72E–06 person-years/kg CO_2. The GWP 100 for NO_x was determined in 6.3.2.2 to –95 ± 31. This means that the environmental impact factor is 1.72E–06*(–95) = –1.63E–04 person-years/kg NO_x.

The uncertainty is the same as for CO_2 (a factor of 2) plus that for the GWP, i.e., a factor 2.1 in a log-normal distribution.

6.3.3.6 Decreased working capacity caused by climate change

The environmental impact factor for CO_2 was determined in 6.1.3.6 to 4.53E–03 person-hours/kg CO_2. The GWP 100 for NO_x was determined in 6.3.2.2 to –95 ± 31. This means that the environmental impact factor is 4.53E–03*(–95) = –4.30E–01 person-hours/kg NO_x.

The uncertainty is the same as for CO_2 (a factor of 2) plus that for the GWP, i.e., a factor 2.1 in a lognormal distribution.

6.3.3.7 Diarrheal diseases caused by climate change and degraded water quality

The environmental impact factor for CO_2 was determined in 6.1.3.7 to 2.69E–10 person-years/kg CO_2. The GWP 100 for NO_x was determined in 6.3.2.2 to –95 ± 31. This means that the environmental impact factor is 2.69E–10*(–95) = –2.56E–08 person-years/kg NO_x.

The same uncertainty is assumed as for YLL from diarrheal diseases and CO_2, i.e., a factor of 3.

6.3.3.8 Decreased crop production capacity caused by oxidants

A best estimate of 90 million tons crop loss per year and an uncertainty factor of 1.5 was estimated in 6.2.3.3.

The global emission of NO_x is 1.22E+11 kg/year (IPCC AR5 WG1, Table 8, SM.19). NO_x contribution to ground level ozone was determined in Table 6.5 to 16.3%.

The environmental impact factor therefore becomes 9E+10*0.163/1.22E+11 = 0.120 kg crop/kg NO_x.

An uncertainty of a factor of 2 is assumed for the environmental impact factor.

6.3.3.9 Decreased crop production capacity caused by climate change

There are two mechanisms by which climate change impacts on crop production capacity. One via altered temperature and precipitation as described in Sec. 6.1.3.8 and one via sea level rise as described in Sec. 6.1.3.9. The environmental impact factor for altered temperature and precipitation was estimated to 1.01E–02 kg crop/kg CO_2. The environmental impact factor for sea level rise was estimated to be 5.29E–03 kg crop/kg CO_2. Added together we obtain an environmental impact factor of 1.54E–02 kg crop/kg CO_2. The GWP 100 for NO_x is –95 ± 31, resulting in an environmental impact factor for NO_x of 1.54E–02*(–95) = –1.46 kg crop/kg NO_x.

The same uncertainty as for CO_2 is assumed, a factor of 3.

6.3.3.10 Decreased meat production capacity caused by climate change

The environmental impact factor for CO_2 was determined in Sec. 6.1.3.10 to be 3.72E–04 kg meat/kg CO_2. The GWP 100 for NO_x is –95 ± 31, resulting in an environmental impact factor for NO_x of 3.72E–04*(–95) = –3.53E–02 kg meat/kg NO_x.

Uncertainty is the same as for crop production capacity, a factor of 3.

6.3.3.11 Decreased fish production capacity caused by eutrophication

According to Diaz and Rosenberg [41], dead zones in sea cover 245 000 km^2. A typical production rate of 10 kg/ha and year will give a total loss of 2.45E+08 kg/year. FAO has recommended that the global catch should level out at 100 million tons for fishing to be sustainable. This corresponds to about 3 kg per ha if all ocean and sea areas are included. Assuming 10 kg/ha for production and fishing at continental shelfs may thus be of a reasonable order.

Global emissions of NO_x as N is estimated to 37.2E+09 kg in AR5 (Table 8, SM.19). According to Galloway et al. [42] the anthropogenic deposition of NO_x to marine areas is 21E+09 kg N/year and 18E+09 kg for NH_3 for the year 1990. About half of the global emission of NO_x is deposited in oceans. The export from rivers to coastal areas is estimated to 47.8E+09 kg N/year [43]. The emissions of BOD is estimated as 0.83 TgNeq/yr. This means that as an average 21/(21 + 18 + 47.8 + 0.83) = 0.24 of the NO_X emitted to air, contribute to dead zones.

The average contribution to decreased fish production capacity from NO_x is therefore 1/(37.2E+09*46/14)*0.24 = 1.96E–12 and the environmental impact factor 2.45E+08*1.96E–12 = 4.80E–04 kg fish/kg NO_x.

The role of dead sea bottoms for fish production capacity is not clarified in this simple model. The average sustainable catch of 3 kg/ha compared to the used value of 10 kg/ha gives an indication of the uncertainty as well as the real variation in contribution. The total uncertainty for the environmental impact indicator is estimated to a factor of 3.

6.3.3.12 Increased fish production capacity caused by fertilization

FAO has recommended that the global catch should level out at 100 million tonnes for fishing to be sustainable. The global deposition of NO_x to oceans is about 40 Tg/year. In the upper 100 m, this will contribute to a 0.123% increase of reactive nitrogen (the ocean area is 360 million km^2, the average concentration of reactive nitrogen is 0.9 g/m^3 and the added concentration from atmospheric deposition is 0.00111 g/m^3 in the upper 100 m). It is here assumed that the increase in fish catch is proportional to nitrogen availability, as nitrogen is rate limiting. This is a rough simplification as many other factors determine which catch is available but gives an idea of a possible magnitude of the impact. The total increase in fish production then becomes 100*0.00111/0.9 = 0.123 million tons or 1.23E+08 kg.

The average contribution from NO_x emissions is the same as for eutrophication (Sec. 6.3.2.11, 1.96E–12) and the environmental impact factor –1.23E+08*1.96E–12 = –2.41E–04 kg fish/kg NO_x.

The uncertainty is estimated to a factor of 4, due to the very primitive impact model.

6.3.3.13 Decreased fish production capacity caused by acidification

Acidification and its impact on fish growth was a topic of great concern during the 1970s and 1980s in northern Europe and northern America. Fish reproduction was not possible when the pH became too low in freshwater. Since then, several measures have been taken to decrease deposition of acid substances from air and to add lime to acidified lakes. Since 1990, the acidification problem has decreased in northern Europe and northern America but increased in other places like southern Africa and Asia. No global estimates have been found for the quantitative consequences of acidification. The impact category "acidification" is currently used in Life Cycle Analysis, why there is a need for at least an estimate of an upper limit for its environmental costs.

An estimation of the impact on fish production capacity can be made from the share of lakes where the critical load for acid deposition are exceeded. Data from Sweden indicates an order of 10% [44] of the lakes although decreasing since 1990s [45]. The acidified lake area in northern Europe and northern America is estimated to be less than 1% of the global lake area subject to fishing. FAO estimated the global freshwater catch to 10 million tons annually. A 1% decrease will correspond to a loss of less than 100, 000 ton of fish annually or 1E+08 kg.

As SO_x also is a cause of lake acidification, the contribution from NO_x should not be 100%. The global yearly emission of SO_2 is 1.27E+05 Gg (AR5, WGII, Table 8 SM 8.19), which results in 1.27E+05/64*2 = 3969 Gmoles of H +, compared to 2660 Gmoles H + from NO_x, from a NO_x emission of 37.2E+09 kg N, and 2447 Gmole from 4.16E+10 kg NH_3.1 kg NO_x then contributes to acidification impacts with 1/3.72E+10*2660/(2660 + 3969 + 2447) = 7.88E–12.

The environmental impact factor thus becomes 1E+08*7.88E–12 = 7.88E–04 kg fish/kg NO_x.

The uncertainty is large, but it is assumed that the real environmental impact factor is less than the estimate.

6.3.3.14 Decreased wood production capacity caused by oxidants

During the 1970s and 1980s damages to forests from air pollution was of great concern. Decline in growth in pine and hardwood was demonstrated at increased ozone concentrations [46]. Decline in growth rates was explained by decreased carbon uptake in leaves and needles. However, a recent review on ozone impacts on European forest have concluded that carbon uptake in leaves and needles is normally not rate limiting for trees growing in forests, which means that the net effect is small or negligible [47].

The impact factor is therefore set to 0.

6.3.3.15 Increased wood production capacity caused by N-fertilization

It has long been assumed that increasing deposition of N from NO_x and NH_3 may increase tree growth, particularly in temperate and boreal forests, where nitrogen is growth limiting. Nadelhoffer et al. [48] reviewed literature estimations of CO_2 sequestration in forests due to nitrogen deposition. Part of the nitrogen is taken up by trees and part of it stays in the soil or leaves the soil in solutions and as gases. They made experiments with adding ^{15}N to a forest area and found that only 5% ended up in the woody biomass. On a global scale they estimated that at most 0.125 Pg CO_2 was sequestrated in woody biomass in forests. 0.125 Pg CO_2 corresponds to 1.25E+11*30/44*1/700*1.2 = 1.46E+08 m^3 wood. The factor 30/44 is the ratio between molecular weights of CH_2O and CO_2 and is used to convert CO_2 uptake to dry wood mass. Traded wood was defined and valued in Chapter 5.4 as wood with 20% moisture. That is why factor 1.2 is used to calculate the wood weight before converting to m^3 by dividing with the density 700 kg/m^3.

The global total emissions of NO_x as N has been estimated to 46 Tg and of NH_3 to 58 Tg [42]. 1 kg of NO_x emitted contributes to 1/(46 + 58)*1E–9 *14/46 = 2.93E–12 of the carbon sequestration. The average environmental impact factor thus becomes –1.46E+08*2.93E–12 = –4.28E–04 m^3 wood/kg NO_x.

The amount of ^{15}N (5%) that was found in wood by Nadelhoffer et al. could have substituted N that would normally had been there. The growth was estimated assuming a constant ratio between C and N. This indicates an uncertainty in the order of a factor of 2.

6.3.3.16 Impacts on wood production capacity via climate change

As for CO_2 (Sec. 6.1.3.12), the net impact on wood production capacity from NO_x is estimated to 0. The GWP 100 for NO_x is –95 ± 31. The environmental impact factor for CO_2 impact on wood production capacity was estimated in Sec. 6.1.3.12 to be 0 ± 6.90E–06 m^3 wood/kg CO_2, why the environmental impact factor for NO_x becomes 0 ± 6.55E–04 kg wood/kg NO_x.

6.3.3.17 Decreased drinking water production capacity via climate change

The environmental impact factor for CO_2 was determined in Sec 6.1.3.13 to 9.06E–04 m^3/kg CO_2 at the most. The GWP 100 for NO_x is –95 ± 31, which means that the environmental impact factor for NO_x is equal to –95*9.06E–04 = –8.61E–02 m^3/kg NO_x.

The uncertainty is high, but as the –8.61E–02 m^3/kg CO_2 is a lower limit, the uncertainty is only towards less negative values.

6.3.3.18 Decreased biodiversity caused by climate change

The environmental impact factor for CO_2 was determined in Sec. 6.1.3.14 to be 1.69E–16 shares of the biodiversity per kg of CO_2. The uncertainty was assumed to be a factor of 4.

The GWP 100 for NO_x is –95 ± 31, which means that the environmental impact factor for NO_x is equal to –95*1.69E–16 = –1.61E–14 shares of the biodiversity/kg NO_x. The same uncertainty factor as for CO_2 is assumed, a factor of 4.

6.3.3.19 Decreased biodiversity caused by eutrophication

The share of species threatened by agricultural and forestry effluents in aquatic environments is 0.007, according to the IUCN Red-list [49]. The contribution of 1 kg of NO_x to loss of fish production capacity in eutrophicated waters was determined in Sec. 6.3.3.11 to be 1.96E–12. The same contribution can be used for the impact on biodiversity. The environmental impact factor then becomes 0.007*1.96E–12 = 1.37E–14 shares of biodiversity per kg of NO_x.

The contribution to eutrophicated waters depends very much on where the emissions occur. An indication of the variations in contributions can be found in the variations in country-specific characterization factors for terrestrial eutrophication as determined by Seppälä et al. [50]. They studied acidification and terrestrial eutrophication in Europe. The characterization factor for NO_2 emissions varied between 0.27 and 3.4 indicating an uncertainty of a factor of 2 from a median value of 1. Two standard deviations (2*2) in a log-normal distribution would give an interval between 0.25 and 4 with a 95% confidence. A factor of 2 is therefore chosen as an uncertainty estimate for the environmental impact factor.

6.3.3.20 Decreased biodiversity caused by acidification

According to IUCN database on red-listed species, 54 species are threatened by acid rain out of the red-listed 113,733 species.

As SO_x and NH_3 also cause lake acidification, the contribution from NO_x should not be 100%. The total global emission of NO_x is 3.72E+04 Gg N (AR5, WGII, Table 8 SM.19), which results in 3.72E+04/14 = 2660 Gmoles H^+. The total global emission of SO_x is 1.27E+05 Gg, which results in 1.27E5/64*2 = 3969 Gmoles H^+. The total global emission of NH_3 is 4.93E+04 Gg, which results in 2447 Gmole of H^+. 1 kg NO_x then contributes with 1/(3.72E+11*46/14)*3969/(2660+3969+2447) = 3.58E–13 to decreased biodiversity.

The environmental impact indicator is thus 54/113733*3.58E–13 = 1.70E–16 shares of the present impact on biodiversity/kg NO_x.

The uncertainty in the numbers used is moderate. An aggregated uncertainty of a factor of 1.5 is assumed.

6.3.4 Summary and monetary valuation of NO_x emissions

A summary of environmental impact factors and monetary values of environmental impacts are shown in Table 6.9.

The highest impact costs are caused by excess mortality from secondary particles and formation of oxidants, mainly ozone. But there is a considerable benefit because NO_x moderates global warming. Excess mortality is decreasing and working capacity increasing as a consequence of NO_x emissions. The benefits are higher than the costs, so the net impact cost of NO_x is negative. Considering all regulations and control measures to decrease NO_x emissions currently applied, it may seem controversial with a measure that says that emissions of NO_x is beneficial to the environment. But there is a plausible explanation for this in the history of environmental concern. Local governments have had the responsibility to tackle local environmental problems. And local environmental problems were the first to be discovered. The negative radiative forcing from NO_x is a recent discovery and appeared first in IPPCs AR5. It is a global issue and global abatement measures take time to negotiate.

Allen et al. [51] have pointed out that if the goal is to decrease the peak temperature from global warming it may not be relevant to use the GWP 100 for short lived climate pollutants, such as NO_x. The GWP 100 is in that case relevant for peak temperatures 20–40 years after the time of emission. In this book, the integrated costs over the entire time of the impact are determined and linear temperature-impact relationships are used.

6.4 Dinitrogen oxide (N_2O)

6.4.1 Specification of the emission

N_2O occurs at low concentrations in combustion gases (a few percent of NO_x). N_2O is also produced from biochemical processes such as denitrification at sewage water treatments and agriculture. Since 1950, N_2O concentration has increased in the atmosphere. Studies of the isotopic composition of N_2O suggest that the increase is mainly caused by emissions from soils treated with synthetic or organic nitrogen fertilizers (IPCC AR5 WG1). It has also been shown that part of the N from NO_x deposited

Table 6.9. Summary of environmental impact factors and monetary impact values of 1 kg of NO_x emissions.

Environmental good	Impact indicator	Unit	Pathway	Environmental impact factor	Uncertainty	Indicator value ($/unit)	Impact value ($/kgNOx)
human health	YLL	personyears	secondary particles	1.96E–06	2	107,067	2.10E–01
human health	YLL	personyears	climate change	–1.79E–04	2	107,067	–1.92E+01
human health	YLL	personyears	oxidant formation	6.97E–06	2	107,067	7.46E–01
human health	YLD	personyears	secondary particles	4.80E–07	2	107,067	5.14E–02
human health	undernutrition	personyears	climate change	–1.63E–04	2.1	6,424	–1.05E+00
human health	working capacity	personhours	climate change	–4.30E–01	2.1	30	–1.29E+01
human health	diarrhea	personyears	climate change	–2.56E–08	3	11,242	–2.88E–04
crop	production capacity	kg	oxidant formation	0.12	2	0.289	3.47E–02
crop	production capacity	kg	climate change	–1.46	3	0.289	–4.22E–01
meat	production capacity	kg	climate change	–3.53E–02	3	2.59	–9.14E–02
fish	production capacity	kg	eutrophication, dead zones	4.80E–04	3	2.42	1.16E–03
fish	production capacity	kg	N-nutrification of ocean	–2.41E–04	4	2.42	–5.83E–04

fish	production capacity	kg	acidification	7.88E–04	upper limit	2.42	1.91E–03
wood	production capacity	m^3	oxidant formation	0		72	0
wood	production capacity	m^3	N-fertilization	–4.28E–04	2	72	–3.08E–02
wood	production capacity	m^3	climate change	0	6.55E–04	72	0
drinking water	production capacity	m^3	climate change	–8.61E–02	upper limit	1.87	–1.61E–01
biodiversity	share of threat to redlisted species	dimensionless	climate change	–1.61E–14	4	7.61E+10	–1.23E–03
biodiversity	share of threat to redlisted species	dimensionless	eutrofication	1.37E–14	2	7.61E+10	1.04E–03
biodiversity	share of threat to redlisted species	dimensionless	acidification	1.70E–16	2	7.61E+10	1.29E–05
						sum	–3.28E+01

on soil is re-emitted in the air as N_2O [52] as well as to oceans [53]. The average life time for N_2O is 121 years, why its impacts are independent of where the emission take place.

The emission, whose environmental damage costs is valued here, is one extra kg of N_2O emitted any place in the world in 2015 and at any source strength.

6.4.2 Impact pathways

Nitrous oxide has no known direct health effect on humans in the concentrations present in ambient air. It is a potent climate gas and contributes to depletion of stratospheric ozone. It is part of the global nitrogen cycle. Most emissions to air from soil and ocean are natural. The total anthropogenic emissions of nitrous oxide are about 6.9 Tg N/year (IPCC AR5 WG1, Chapter 6). The natural emissions are about 11.0 Tg N/year. Of the anthropogenic emissions, burning of fossil fuels and biomass contributes with 0.7 Tg N each and agriculture with 4.1 from fertilizers. 0.6 Tg N is re-emitted N deposition from air.

All pathways except one include effects of global warming and results in the same types of effects as carbon dioxide. There is a slight difference in that carbon dioxide has one type of fertilizing effect and nitrous oxide another, but this is assumed to be small and is not modelled here. The pathways modeled are shown in Table 6.10.

Table 6.10. Pathways by which nitrous oxide, N_2O, impact on environmental goods.

Environmental good	Impact indicator	Unit	Pathway
human health	YLL	personyears	climate change
human health	YLL	personyears	ozone depletion
human health	undernutrition	personyears	climate change
human health	working capacity	personhours	climate change
human health	diarrhea	personyears	climate change
crop	production capacity	kg	climate change
meat	production capacity	kg	climate change
fish	production capacity	kg	climate change
wood	production capacity	m^3	climate change
drinking water	production capacity	m^3	climate change
biodiversity	share of threat to redlisted species	dimensionless	climate change

6.4.3 Environmental impact factors

6.4.3.1 YLL via climate change

N_2O has a global warming potential of 265. This means that the extent of the impacts on YLL is the same as for CO_2, 1.11E+10 YLL for the period 2015–2100, but the contributions from 1 extra kg of N_2O is 265 times higher, i.e., 4.48E–14 per kg.

The environmental impact factor is then 1.11E+10*4.48E–14 = 4.97E–04 YLL/kg N_2O.

The uncertainty is the same as for the highest contribution to YLL for CO_2 (undernutrition), a factor of 2.

6.4.3.2 YLL via stratospheric ozone depletion

The emission scenario based on the Copenhagen amendment to the Montreal Protocol is used. For that, WHO estimates an average incidence of around 50 cases per million inhabitants per year until 2100, when excess cancer incidence cease. There are three types of skin cancer: melanoma, squamous cell carcinoma (SCC), and basal cell carcinoma (BCC). Mortality in skin cancer is low, about 12% for melanoma and 0,2% for SCC and BCC. If applying the ratio of predicted cancer incidences for melanoma vs. SCC + BCC reported to USEPA [54], where the melanoma incidence is estimated to be 2.7% of all cases, the global mortality rate until 2100 will be 50/1E+06*7.2E9*88*(0.12*0.027 + 0.002*0.973) = 1.64E+05.

The average shortening of life due to cancer is 24 years. This means that there are 3.94E+06 YLL caused by ozone depletion.

The total emission of ozone depletion substances to 2100 is estimated from a scenario made by A.R. Ravishankara et al. [55] to 2.2E+10 kg CFC11 equivalents. An emission of 1 kg CFC11 equivalents therefore contributes with 3.94E+06/2.2E+10 = 1.79E–4 YLL/kg. N_2O has an ozone depletion potential of 0.017 [55]. The environmental impact factor for N_2O is therefore 0.017*1.79E–4 = 3.04E–06 YLL/kg N_2O.

There is significant uncertainty in the exposure scenario and in adaptation. An uncertainty of a factor of 4 is assumed.

6.4.3.3 Increased undernutrition via climate change

N_2O has a global warming potential of 265. This means that the extent of the impacts on undernutrition is the same as for CO_2, 1.02E+10 person-years for the period 2015–2100, but the contributions from 1 extra kg of N_2O is 265 times higher, i.e., 4.48E–14 per kg.

The environmental impact factor is then 1.02E+10*4.48E–14 = 4.57E–04 person-years/kg N_2O.

The uncertainty is the same as for CO_2, a factor of 4.

6.4.3.4 Decreased working capacity via climate change

N_2O has a global warming potential of 265. This means that the extent of the impacts on working capacity is the same as for CO_2, 2.68E+13 person-hours for the period 2015–2100, but the contributions from 1 extra kg of N_2O is 265 times higher, i.e., 4.48E–14 per kg.

The environmental impact factor is then 2.68E+13*4.48E–14 = 4.57E–04 person-hours/kg N_2O.

The uncertainty is the same as for CO_2, a factor of 4.

6.4.3.5 Increased diarrhea via climate change and polluted drinking water

N_2O has a global warming potential of 265. This means that the extent of the impacts on diarrheal diseases is the same as for CO_2, 1.59E+06 person-years for the period 2015–2100, but the contributions from 1 extra kg of N_2O is 265 times higher, i.e., 4.48E–14 per kg.

The environmental impact factor is then 1.59E+06*4.48E–14 = 7.12E–08 person-years/kg N_2O.

The uncertainty is the same as for CO_2, a factor of 3.

6.4.3.6 Decreased crop production capacity via climate change

N_2O has a global warming potential of 265. This means that the extent of the impacts on crop production capacity is the same as for CO_2, 9.12E+13 kg crop for the period 2015–2100, but the contributions from 1 extra kg of N_2O is 265 times higher, i.e., 4.48E–14 per kg.

The environmental impact factor is then 9.12E+13*4.48E–14 = 4.08 kg crop/kg N_2O.

The uncertainty is the same as for CO_2, a factor of 3.

6.4.3.7 Decreased meat production capacity via climate change

N_2O has a global warming potential of 265. This means that the extent of the impacts on meat production capacity is the same as for CO_2, 2.20E+12 kg meat for the period 2015–2100, but the contributions from 1 extra kg of N_2O is 265 times higher, i.e., 4.48E–14 per kg.

The environmental impact factor is then 2.20E+12*4.48E–14 = 9.85E–02 kg meat/kg N_2O.

The uncertainty is the same as for CO_2, a factor of 3.

6.4.3.8 Decreased fish production capacity via climate change

N_2O has a global warming potential of 265. This means that the extent of the impacts on fish production capacity is the same as for CO_2, 1.73E+11 kg fish for the period 2015–2100, but the contributions from 1 extra kg of N_2O is 265 times higher, i.e., 4.48E–14 per kg.

The environmental impact factor is then 1.73E+11*4.48E–14 = 7.75E–03 kg fish/kg N_2O.

The uncertainty is the same as for CO_2, a factor of 2.

6.4.3.9 Impact on wood production capacity via climate change

N_2O has a global warming potential of 265. This means that the net extent of the impacts on wood production capacity is the same as for CO_2, 0 m^3 wood for the period 2015–2100, but the contributions from 1 extra kg of N_2O is 265 times higher, i.e., 4.48E–14 per kg.

The environmental impact factor is then 0*4.48E–14 = 0 m^3 wood/kg N_2O.

The uncertainty is the same as for CO_2, but multiplied with the GWP of 265, i.e., ± 1.83E–03 m^3 wood/kg N_2O.

6.4.3.10 Decreased drinking water production capacity via climate change

N_2O has a global warming potential of 265. This means that the extent of the impacts on drinking water production capacity is the same as for CO_2, 5.36E+12 m^3 for the period 2015–2100, but the contributions from 1 extra kg of N_2O is 265 times higher, i.e., 4.48E–14 per kg.

The environmental impact factor is then 5.36E+12*4.48E–14 = 2.40E–01 m^3 drinking water/kg N_2O.

The uncertainty is the same as for CO_2, a factor of 3.

6.4.3.11 Decreased biodiversity via climate change

N_2O has a global warming potential of 265. This means that the extent of the impacts on biodiversity is the same as for CO_2, 1, for the period 2015–2100, but the contributions from 1 extra kg of N_2O is 265 times higher, i.e., 4.48E–14 per kg.

The environmental impact factor is then 1 *4.48E–14 = 4.48E–14 shares of present threat to biodiversity/kg N_2O.

The uncertainty is the same as for CO_2, a factor of 4.

Table 6.11. Summary of environmental impact factors and monetary impact values of emission of 1 kg N_2O.

Environmental good	Impact indicator	Unit	Pathway	Environmental impact factor	Uncertainty	Indicator value ($/unit)	Impact value ($/kgN$_2$ CO)
human health	YLL	personyears	climate change	4.99E–04	2	107067	5.35E–01
human health	YLL	personyears	ozone depletion	3.04E–06	4	107067	3.26E–01
human health	undernutrition	personyears	climate change	4.57E–04	4	6424	2.93E–00
human health	working capacity	personhours	climate change	1.20E–00	4	30	3.60E+01
human health	diarrhea	personyears	climate change	7.12E-08	3	11242	8.00E–04
crop	production capacity	kg	climate change	4.08E+00	3	0.289	1.18E–00
meat	production capacity	kg	climate change	9.85E–02	3	2.59	2.55E–01
fish	production capacity	kg	climate change	7.75E–03	2	2.42	1.88E–02
wood	production capacity	m^3	climate change	0	1.83E–03	72	0
drinking water	production capacity	m^3	climate change	2.40E–01	3	1.87	4.49E–01
biodiversity	share of threat to redlisted species	dimensionless	climate change	4.48E–14	4	7.61E+10	3.41E–03
						Sum	9.46E+01

6.4.4 Summary and monetary valuation of N_2O emissions

Table 6.11 gives the summary of environmental impact factors and monetary impact values of emission of 1 kg N_2O.

As for CO_2, the highest impact values are from reduced life expectancy and reduced working capacity.

6.5 Ammonia

6.5.1 Specification of the emission

The main source of NH_3 is agriculture, generating 30.4 TgN per year. Biomass and biofuel burning generates 9.2 TgN. Natural sources generate 10.6 TgN. The total global emission to air is 50.7 TgN per year (IPPC AR5 WG1, Chapter 6).

The emission valued is an emission anywhere in the world during the year 2015.

6.5.2 Impacts caused on environmental goods and impact pathways

NH_3 has a short residence time in air. It reacts with acid sulfates and nitrates to form ammonia sulfates and nitrates. The residence time for ammonia and NH_4^+, is about a week, and that is why the effects are local to regional.

The impacts caused by NH_3 is mainly via its secondary particles and via fertilizing. The impacts modelled here are shown in Table 6.12.

6.5.3 Environmental impact factors

6.5.3.1 YLL via secondary aerosols

Lelieveld et al. [20] calculated the world excess mortality from PM2.5 to 1.59E+07 YLL for the year 2005.

The mass fraction of NH_4^+ in PM2.5 in ambient air is around 4% (IPCC AR5 WG1, Chapter 7) and the global emission of NH_3 is 4.16E+10 kg/year (IPCC AR5 WG1 Table 8, SM.19). Assuming much lower health effects for NH_4^+ than the average PM2.5 (25%), the contribution to the 1.59E+07 YLL/year is 0.04*(4.16E+10) = 2.40E–13 and the environmental impact factor 1.59E+07*2.40E–13 = 3.82E–6 YLL/kg NH_3.

There is an uncertainty in the dose-response function. A factor of 2 is assumed and would imply that the effect is between 6 and 100% with 95% probability.

Table 6.12. Pathways by which NH_3 impacts on environmental goods.

Environmental good	Impact indicator	Unit	Pathway
human health	YLL	personyears	seconday aerosols
human health	YLL	personyears	climate change
human health	YLD	personyears	seconday aerosols
human health	undernutrition	personyears	climate change
human health	working capacity	personyears	climate change
human health	diarrhea	personyears	climate change
crop	production capacity	kg	climate change
meat	production capacity	kg	climate change
fish	production capacity	kg	acidification
fish	production capacity	kg	eutrophication
fish	production capacity	kg	fertilizing
wood	production capacity	m^3	fertilizing
drinking water	production capacity	m^3	climate change
biodiversity	share of threat to redlisted species	dimensionless	acidification
biodiversity	share of threat to redlisted species	dimensionless	climate change
biodiversity	share of threat to redlisted species	dimensionless	eutrophication

6.5.3.2 YLL via climate change

GWP100 for NH_4^+ particles is not presented in IPCC AR5 WG1, but should be the same as for sulfate particles, –40 with respect to SO_2 [56]. The corresponding GWP100 for NH_3 is then 64/17*(–40) = –151. Practically all the NH_3 emitted is transformed to particles. The global excess mortality for CO_2 was determined in 6.1 (Table 6.3). The sum of all environmental impact factors for YLL is 1.88E–06 YLL/kg CO_2. This means that the environmental impact factor for NH_3's impact on YLL via climate change is 1.88E–06*(–151) = –2.84E–04 YLL/kg NH_3.

NH_3 and NH_4 particles have a short residence time in air, about a week. GWP100 for NH_3 is likely to vary between regions with the same magnitude as GWP 100 for NO_x, i.e., a factor of 2 from an average (IPCC AR5 WG1 Table 8.A.3). The uncertainty of the environmental impact factor for CO_2 and YLL is estimated to a factor of 2, which is the uncertainty of the highest environmental impact factor (undernutrition). The total uncertainty factor is then $\exp\sqrt{(\ln 2)^2 + (\ln 2)^2} = 2.7$.

6.5.3.3 YLD through cardiovascular diseases caused by secondary aerosols

Cohen et al. [40] found that long-term exposure to PM2.5 caused 3.9 million (2.6 million to 5.2 million) years lived with disability (YLD), in 2015.

The mass fraction of NH_4^+ is only around 4% (IPCC AR5 WG1, Chapter 7) and the global emission of NH_3 is 4.16E+10 kg/year (IPCC AR5 WG1 Table 8, SM.19). Assuming a much lower health effect for NH_4^+ than the average PM2.5 (25%), the contribution to the 3.9E+06 YLD/year is 0.04*0.25/(4.16E+10) = 2.40E–13 and the environmental impact factor 3.9E+06*2.40E–13 = 9.36E–07 YLD/kg NH_3.

The uncertainty given by Cohen et al. concerns PM2.5. There is also an uncertainty in the estimation of the toxicity of NH_4^+ and in the contribution from NH_3-emissions. A factor of 2 is assumed for the standard deviation of the environmental impact factor.

6.5.3.4 Undernutrition via climate change

For CO_2, the environmental impact factor was determined in 6.1.3.5 to 1.72E–06 person-years/kg CO_2. The GWP 100 for NH_3 was determined in 6.5.3.2 to –151. This means that the environmental impact factor is 1.72E–06*(–151) = –2.60E–04 person-years/kg NO_x.

The uncertainty is the same as for CO_2 (a factor of 2) plus that for the GWP, i.e., a factor 2.1 in a log-normal distribution.

6.5.3.5 Decreased working capacity via climate change

The environmental impact factor for CO_2 was determined in 6.1.3.6 to 4.53E–03 person-hours/kg CO_2. The GWP 100 for NH_3 was determined in 6.5.3.2 to –151. This means that the environmental impact factor is 4.53E–03*(–151) = –6.84E–01 person-hours/kg NH_3.

The uncertainty is the same as for CO_2 (a factor of 2) plus that for the GWP, i.e., a factor of 2.1 in a log-normal distribution.

6.5.3.6 Increased diarrhea via climate change and polluted drinking water

The environmental impact factor for CO_2 was determined in Sec. 6.1.3.7 to 2.69E–10 person-years/kg CO_2.The GWP 100 for NH_3 was determined in 6.5.3.2 to –151. This means that the environmental impact factor is 2.69E–10*(–151) = –4.06E–08 person-years/kg NH_3.

The same uncertainty is assumed as for YLL from diarrheal diseases and CO_2, i.e., a factor of 3.

6.5.3.7 Decreased crop production capacity via climate change

There are two mechanisms by which climate change impacts on crop production capacity. One via altered temperature and precipitation as described in 6.1.3.8 and one via sea level rise as described in 6.1.3.9. The environmental impact factor for altered temperature and precipitation was estimated to 1.01E–02 kg crop/kg CO_2. The environmental impact factor for sea level rise was estimated to 5.29E–03 kg crop/kg CO_2. Added together an environmental impact factor of 1.54E–02 kg crop/kg CO_2 is obtained. The GWP100 for NH_3 was determined in 6.5.3.2 to –151, resulting in an environmental impact factor for NH_3 of 1.54E–02*(–151) = –2.33 kg crop/kg NH_3.

The same uncertainty as for CO_2 is assumed, a factor of 3.

6.5.3.8 Decreased production capacity for meat via climate change

The environmental impact factor for CO_2 was determined in 6.1.3.10 to 3.72E–04 kg meat/kg CO_2. The GWP 100 for NH_3 was determined in 6.5.3.2 to –151, resulting in an environmental impact factor for NH_3 of 3.72E–04*(–151) = –5.62E–02 kg meat/kg NO_x.

Uncertainty is the same as for crop production capacity, a factor 3.

6.5.3.9 Decreased fish production capacity caused by acidification

It may seem odd that NH_3 is an acidifying substance. Initially it neutralizes acidity, but when oxidized it will form HNO_3 and contribute to acidification.

An estimation of the impact on fish production capacity can be made from the share of lakes where the critical load for acid deposition are exceeded. Data from Sweden indicates an order of 10% [44] of the lakes although decreasing since 1990s [45]. The acidified lake area in northern Europe and northern America is estimated to be less than 1% of the global lake area subject to fishing. FAO estimated the global freshwater catch to be 10 million tons annually. A 1% loss will correspond to 100000 ton of fish annually or 1E+08 kg.

As NO_x and SO_x also cause lake acidification, the contribution from NH_3 should not be 100%. The global emission of SO_x is 1.27E+05 Gg (AR5, WGII, Table 8 SM.19), which results in 1.27E+05/64*2 = 3969 Gmoles of H+, compared to the = 2660 Gmoles H + from NO_x, from a NO_x emission of 37.2E+09 kg N, and 2447 Gmole from 4.16E+10 kg NH_3/year. An approximate estimate is that 1 kg NH_3 contributes with 1/4.16E+10*2447/(2660 + 3969 + 2447) = 6.48E–12.

The environmental impact factor thus becomes less than 1E+08* 6.48E–12 = 6.48E–04 kg fish per kg NH_3.

The uncertainty is large, but it is assumed that the real environmental impact factor is less than the estimate.

6.5.3.10 Decreased fish production capacity from eutrophication

According to Diaz and Rosenberg [41], dead zones in sea cover 245 000 km^2. A typical production rate of 10 kg fish/ha and year will give a total loss of 2.45E+08 kg/year. FAO has recommended that the global catch should level out at 100 million tonnes for fishing to be sustainable. This corresponds to about 3 kg per ha if all ocean and sea areas are included. Assuming 10 kg/ha for production and fishing at continental shelves may thus be of a reasonable order.

Global emissions of NH_3 is estimated to be 4.16E+10 kg/year in IPCC AR5 (Table 8 SM.19). According to Galloway et al. [42], the anthropogenic deposition of NO_x to marine areas was 21E+09 kg N/year and 18E+09 kg for NH_3 for the year 1990. About half of the global emission of NO_x is deposited in oceans. The export from rivers to coastal areas is estimated to be 47.8E+09 kg N/year [43]. The emissions of BOD are estimated to be 0.83 TgNeq/yr. This means that as an average 18/(21 + 18 + 47.8 + 0.83) = 0.21 of the NH_3 emitted to air, contribute to dead zones.

The average contribution to decreased fish production capacity from NH_3 is therefore 1/(41.6E+09)*0.21 = 5.05E–12 and the environmental impact factor 2.45E+08*5.05E–12 = 1.24E–03 kg fish/kg NH_3.

The role of dead sea bottoms for fish production capacity is not clarified in this simple model. The average sustainable catch of 3 kg/ha compared to the used value of 10 kg/ha gives an indication of the uncertainty as well as the real variation in contribution. The total uncertainty for the environmental impact indicator is estimated to a factor of 3.

6.5.3.11 Increased fish production capacity caused by N fertilizing

FAO has recommended that the global catch should level out at 100 million tons for fishing to be sustainable. The global deposition of NO_x to oceans is about 40 Tg/year. In the upper 100 m:s this will contribute to a 0.123% increase of reactive nitrogen (the ocean area is 360 million km^2, the average concentration of reactive nitrogen is 0.9 g/m^3 and the added concentration from atmospheric deposition is 0.00111 g/m^3 in the upper 100 m). It is here assumed that the increase in fish catch is proportional to nitrogen availability, as nitrogen is rate limiting. This is a rough simplification as many other factors determine which catch is available, but it gives an idea of a possible magnitude of the impact. The total increase in fish production then becomes 100*0.00111/0.9 = 0.123 million tons or 1.23E+08 kg.

Global emissions of NH_3 is estimated to be 4.16E+10 kg/year in IPCC AR5 (Table 8 SM.19). According to Galloway et al. [42], the deposition of anthropogenic NH_3tomarine areas is 18E+09 kg N and for NO_x 21E+09 kg N/year and for the year 1990. About half of the global emission of NO_x is deposited in oceans. The export from rivers to coastal areas is estimated to be 47.8E+09 kg N/year [43]. The emissions of BOD are estimated to be 0.83 TgNeq/yr. This means that as an average 18/(21 + 18 + 47.8 + 0.83) = 0.21 of the NH_3 emitted to air, contribute to dead zones.

The environmental impact factor is then –1.23E+08*1.96E–12 = –2.41E–04 kg fish/kg NH_3.

The uncertainty is estimated to a factor of 4, due to the very primitive impact model.

6.5.3.12 Increased wood production capacity via N fertilization

It has long been assumed that increasing deposition of N from NO_x and NH_3 may increase tree growth, particularly in temperate and boreal forests, where nitrogen is growth limiting. Nadelhoffer et al. [48] reviewed literature estimations of CO_2 sequestration in forests. Part of the nitrogen is taken up by trees and part of it is stays in the soil or leaves the soil in solutions and as gases. They made experiments with adding ^{15}N to a forest area and found that only 5% ended up in the woody biomass. On a global scale they estimated that at most, 0.125 Pg CO_2 was sequestrated in woody biomass in forests. 0.125 Pg CO_2 corresponds to 1.25E+11*30/44*1/700*1.2 = 1.46E+08 m^3 wood. The factor 30/44 is the ratio between molecular weights of CH_2O and CO_2 and is used to convert CO_2 uptake to dry wood mass. Traded wood was defined and valued in Chapter 5.4 as wood with 20 % moisture. That is why factor 1.2 is used to calculate the wood weight before converting to m^3 by dividing with the density 700 kg/m^3.

The global total emissions of NH_3 has been estimated as 58 Tg and of NO_x to N 46 Tg [42]. 1 kg of NH_3 emitted contributes to 1/(46 + 58)*1E–9 *14/17 = 7.92E–12 of the carbon sequestration. The average environmental impact factor thus becomes –1.46E+08*7.92E–12 = –1.16E–03 m^3 wood/kg NO_x.

Nadelhoffer et al. mention that the amount of ^{15}N (5%) that was found in the wood could have substituted N that would normally had been there which would mean that the environmental impact factor would be lower. However, they did not measure growth per se. The growth was estimated assuming a constant ratio between C and N. This indicates an uncertainty in the order of a factor of 2.

6.5.3.13 Decreased drinking water production capacity via climate change

The environmental impact factor for CO_2 was determined in 6.1.3.13 to 9.06E–04 $m^3/kg\ CO_2$ at most. The GWP 100 for NH_3 was determined in 6.5.3.2 to –151, which means that the environmental impact factor for NH_3 is equal to –151*9.06E–04 = –1.37E–01 $m^3/kg\ NH_3$.

The uncertainty is high, but as the –1.37E–01 $m^3/kg\ CO_2$ is a lower limit, the uncertainty is only towards less negative values.

6.5.3.14 Decreased biodiversity via acidification

According to IUCN database on red listed species, 54 species are threatened by acid rain out of red listed 113733 species. The anthropogenic emissions of SO_2 for 2010 has been estimated to 100 Tg by Klimont et al. [57].

As SO_x and NO_x also cause lake acidification, the contribution from NH_3 should not be 100%. The total global emission of SO_x is 1.27E+05 Gg (AR5, WGII, Table SM 8.19), which results in 1.27E+05/64*2 = 3969 Gmoles of H+, compared to the 2660 Gmoles of H+ from NO_x, from a NO_x emission of 37.2E+09 kg N, and 2447 Gmole from 4.16E+10 kg NH_3/year. 1 kg NH_3then contributes with 1/4.16E+10*2447/(2660 + 3969 + 2447) = 6.48E–12 to decreased biodiversity.

The environmental impact indicator is thus 54/113733*6.48E–12 = 3.08E–15 shares of the present impact on biodiversity/kg NH_3.

The uncertainty in the numbers used is moderate. An aggregated uncertainty of a factor of 1.5 is assumed.

6.5.3.15 Decreased biodiversity via climate change

The environmental impact factor for CO_2 was determined in 6.1.3.14 to 1.69E–16 shares of the biodiversity per kg of CO_2. The uncertainty was assumed to be a factor of 4.

The GWP 100 for NH_3 was determined in 6.5.3.2 to –151, which means that the environmental impact factor for NH_3 is equal to –151*1.69E–16 = –2.55E–14 shares of the biodiversity/kg NH_3. The same uncertainty factor as for CO_2 is assumed, a factor of 4.

6.5.3.16 Decreased biodiversity caused by eutrophication

The share of species threatened by agricultural and forestry effluents in aquatic environments is 0.007, according to the IUCN Redlist [49]. The contribution of 1 kg of NO_x to loss of fish production capacity in eutrophicated waters was determined in 6.4.3.11 to 5.05E–12. The same contribution can be used for the impact on biodiversity. The environmental impact factor then becomes 0.007*5.05E–12 = 3.54E–14 shares of biodiversity per kg of NH_3.

The contribution to eutrophicated waters depends very much on where the emissions occur. An indication of the variations in contributions can be found in the variations in country-specific characterization factors for terrestrial eutrophication as determined by Seppälä et al. [50]. They studied acidification and terrestrial eutrophication in Europe. The characterization factor for NO_2 emissions varied between 0.27 to 3.4 indicating an uncertainty of a factor of 2 from a median value of 1. Two standard deviations (2*2) in a log-normal distribution would give an interval between 0.25 and 4 for a 95% confidence. A factor of 2 is therefore chosen as an uncertainty estimate for the environmental impact factor.

6.5.4 Summary and monetary valuation of NH_3 emissions

A summary of estimated environmental impact factors and monetary impact values are shown in Table 6.13.

6.6 Sulfur Oxides

6.6.1 Specification of the emission

Sulfur oxides, mostly SO_2 and a few % SO_3, are emitted from anthropogenic sources, such as energy production sites, smelters and pulp mills. Volcanoes, forest fires and bio-processes generating H_2S, COS, DMS, and CH_3SH, which are oxidized in air to SO_2, are examples of natural sources.

The globally emitted SO_2 is about 79 Tg S/year of which 82% of this is anthropogenic according to Rash et al. [58]. Their estimations were made from a global model. The main sources of sulfur were SO_2 and DMS. Lee et al. [59] estimated the SO_2 formed from oxidation of DMS to 31 Tg S. The residence time of sulfur dioxide in the atmosphere is short, in the order of 1–2 days. SO_2 is soon oxidized to sulfate mostly in clouds, but also in the gas phase. The life time of sulfate particles is 4–5 days, and most of the particles are removed from the atmosphere through wet deposition.

This means that the location of a sulfur dioxide source has a great influence on its impacts, and that lack of spatial information adds to the uncertainty of its impacts.

The emissions valued are sulfur oxides, SO_2 + SO_3, emitted anywhere in the world 2015. The unit measure refers to SO_2.

Table 6.13. Summary of environmental impact factors and monetary impact values of NH_3 emissions.

Environmental good	Impact indicator	Unit	Pathway	Environmental impact factor	Uncertainty	Indicator value ($/unit)	Impact value ($/kg NH_3)
human health	YLL	personyears	seconday aerosols	3.82E–06	2	107067	4.09E–01
human health	YLL	personyears	climate change	–2.84E–04	2.8	107067	3.04E+01
human health	YLD	personyears	seconday aerosols	9.36E–07	2	107067	1.00E–01
human health	undernutrition	personyears	climate change	–2.60E–04	2.1	6424	–1.67E+00
human health	working capacity	personyears	climate change	–6.84E–01	2.1	30	–2.05E+01
human health	diarrhea	personyears	climate change	–4.06E–08	3	11242	–4.57E–04
crop	production capacity	kg	climate change	–2.33	3	0.289	–6.73E–01
meat	production capacity	kg	climate change	–5.62E–02	3	2.59	–1.45E–01
fish	production capacity	kg	acidification	6.48E–04	upper limit	2.42	1.57E–03
fish	production capacity	kg	eutrophication	1.24E–03	3	2.42	3.00E–03
fish	production capacity	kg	fertilizing	–2.41E–04	4	2.42	–5.83E–04
wood	production capacity	m^3	fertilizing	–1.16E–03	2	72	–8.35E–02
drinking water	production capacity	m^3	climate change	–1.37E–01	upper limit	1.87	–2.56E–01
biodiversity	share of threat to redlisted species	dimensionless	acidification	3.08E–15	1.5	7.61E+10	2.34E–04
biodiversity	share of threat to redlisted species	dimensionless	climate change	–2.55E–14	4	7.61E+10	–1.94E–03
biodiversity	share of threat to redlisted species	dimensionless	eutrophication	3.54E–14	2	7.61E+10	2.69E–03
						sum	–5.32E+01

6.6.2 Impacts caused on environmental goods and impact pathways

Sulfur dioxide has direct and indirect health effects on the respiratory system, it reacts in air to produce sulfate particles which influence climate, it is a fertilizer, it contributes to acidification and corrosion of materials. Its fertilizing effect is not modeled here as it is thought to minor significance. The impacts modeled are shown in Table 6.14.

Table 6.14. Pathways by which sulfur oxides, SO_x, impact on environmental goods.

Environmental good	Impact indicator	Unit	Pathway
human health	YLL	personyears	direct exposure
human health	YLL	personyears	secondary particles
human health	YLL	personyears	climate change
human health	YLD	personyears	direct exposure
human health	YLD	personyears	secondary particles
human health	undernutrition	personyears	climate change
human health	working capacity	personhours	climate change
human health	diarrhea	personyears	climate change
crop	production capcity	kg	climate change
meat	production capcity	kg	climate change
fish	production capcity	kg	acidification
wood	production capcity	m^3	climate change
drinking water	production capcity	m^3	climate change
biodiversity	share of threat to redlisted species	dimensionless	climate change
biodiversity	share of threat to redlisted species	dimensionless	acidification
all	CO_2 emission	kg	corrosion

6.6.3 Environmental impact factors

6.6.3.1 Impacts on YLL via direct exposure

The dose-response for mortality due to SO_2 exposure is estimated to be + 0.6% per 10 μg/m^3 (24 hour mean) [60]. No threshold is found, so it is assumed here that the dose-response curve is linear, why it is applicable also to yearly averages. The average annual urban population exposure to SO_2 was determined to 30 μg/m^3 for the European region by WHO regional office 1997. Between 2002 to 2011, EU27 member states decreased their SO_x emissions by 50% and the air concentration by 30% [61]. The highest concentrations nowadays are found in India and China, but the pattern changes fast [62].

Lacking new estimations of the global average SO_2 concentration in ambient air, the old European estimate, 30 μg/m³ will be used for the present global, urban population, which was 51,6% of the total world population in 2010. The lost life expectancy per case of premature death is estimated to be the same as for heat stroke, 1 month, as it has a similar health effect for persons with cardiovascular diseases. The total YLLs due to SO_2-exposure is therefore estimated to be 0.006*30/10*0.516*7.4E+09/72 = 9.55E+05 YLL, where 72 is an estimated average life expectancy and 7.4E+09 is the world population 2015.

Although much of the global SO_x originates from natural sources, primarily oxidation of DMS, population exposure is dominated by anthropogenic emissions. The anthropogenic emissions of SO_2 for 2010 has been estimated to 100 Tg by Klimont et al. [57]. The contribution to the impact is thus 1E–11 per kg SO_2. The environmental impact indicator is 9.55E+05*1E–11 = 9.55E–06 YLL/kg SO_2.

No recent estimation of the global average concentration of SO_2 in ambient air is found, why the population exposure is uncertain, but should be within double and half the values. An uncertainty of a factor of 2 is assumed.

6.6.3.2 *Impacts on YLL via exposure to secondary particles*

As mentioned in Sec. 6.3.3.1, there is an uncertainty if and how much sulfate and nitrate particles contribute to the health effects of PM2.5. The same assumption is made here for sulfates as was made for nitrates, i.e., that their toxicity is 25% of the average PM2.5.

Lelieveld et al. [20] calculated the world excess mortality from PM2.5 to 1.59E+07 YLL for the year 2005.

The mass fraction of sulfates normally varies between 10 and 30% (IPCC AR5 WG1, Chapter 7) and the global anthropogenic emission of SO_2 is 1E+11 kg/year [57]. Assuming a much lower health effects for sulfates than the average PM2.5 (25%), the contribution to the 1.59E+07 YLL/year is 0.15*0.25/(1E+11) = 3.75E–13 and the environmental impact factor 1.59E+07*3.75E–13 = 5.96E–06YLL/kg SO_x.

There is an uncertainty in the exposure and the dose-response function. A factor of 2 is assumed for that uncertainty. There is also an uncertainty in the contribution factor. SO_2 has a short residence time in air and regional variations occur. A factor of 3 is estimated for the total uncertainty.

6.6.3.3 Impacts on YLL via climate change

GWP100 for sulfate particles is –40 [56]. About half of the SO_2 emitted is transformed to sulfate particles [63]. This gives a particle formation potential of 0.5*96/64 = 0.75. The global excess mortality 2015–2100 was determined in Sec. 6.1 (Table 6.3). The sum of all environmental impact factors for YLL is 1.88E–06 YLL/kg CO_2. This means that the environmental impact factor for SO_2's impact on YLL via climate change is 1.88E–06*(–40)*0.75 = –5.64E–05 YLL/kg SO_2.

SO_x and sulfate particles have a short residence time in air, about a week. GWP100 for SO_2 is likely to vary between regions with the same magnitude as GWP 100 for NO_x, i.e., a factor of 2 from an average (IPCC AR5 WG1 Table 8 A.3). The uncertainty of the environmental impact factor for CO_2 and YLL is estimated to a factor of 2, which is the uncertainty of the highest environmental impact factor (undernutrition). The total uncertainty factor is then exp $\sqrt{(\ln 2)^2 + (\ln 2)^2} = 2.7$.

6.6.3.4 Impacts on YLD via direct exposure

The dose-response for respiratory hospital admissions due to SO_2 exposure is estimated as + 0.5% per 10 μg/m3 (24 hour mean) [60]; No threshold is found so it is assumed here that the dose-response curve is linear, why it is applicable also to yearly averages. Globally there are about 50 million YLD from respiratory diseases [13], mostly being COPD and asthma. The average annual urban population exposure to SO_2 was determined to 30 μg/m^3 for the European region by WHO regional office 1997. Between 2002 and 2011 EU27 member states decreased their SO_x emissions by 50% and the air concentration by 30% [61]. The highest concentrations nowadays are found in India and China, but the pattern changes fast [62]. Lacking new estimations of the global average SO_2 concentration in ambient air, the old European estimate, 30 μg/m^3, will be used for the global average. The total YLDs due to SO_2-exposure is therefore estimated to be 0.005*30/10*50E+06 = 7.50E+05 YLD.

Although much of the global SO_x originates from natural sources, primarily oxidation of DMS, population exposure is dominated by anthropogenic emissions. The anthropogenic emissions of SO_2 for 2010 has been estimated to 100 Tg by Klimont et al. [57]. The contribution to the impact is thus 1E–11 per kg SO_2. The environmental impact indicator is 7.50E+05*1E–11 = 7.50E–06 YLD/kg SO_2.

No recent estimation of the global average concentration of SO_2 in ambient air is found, why the population exposure is uncertain, but should be within double and half the values. An uncertainty of a factor of 2 is assumed.

6.6.3.5 Impacts on YLD via exposure to secondary particles and respiratory diseases

There is no strong indications of health effects of sulfate and nitrate particles in air [35, 64]. In 6.3.3.1 the risk for YLL from respiratory diseases was estimated to be 25% of that of PM2.5. PM2.5 caused 3·9 million YLD (2·6 million to 5·2 million) globally 2015 [40].

The mass fraction of sulfates normally varies between 10 and 30% (IPCC AR5 WG1, Chapter 7) and the global anthropogenic emission of SO_2 is 1E+11 kg/year [57]. The contribution to the 3.9 million YLD is 0.15*0.25/(1E+11) = 3.75E–13 and the environmental impact factor 3.9E+06*3.75E–13 = 1.46E–06YLD/kg SO_x.

The uncertainty in the extent of the impact is about a factor of 1.4. The uncertainty of the contribution estimate is a factor of 2, due to regional variations. The added uncertainty of the impact factor is estimated to a factor of 2.2.

6.6.3.6 Impacts on undernutrition via climate change

For CO_2, the environmental impact factor for undernutrition was determined in Sec. 6.1.3.5 to 1.72E–06 person-years/kg CO_2. The GWP 100 for SO_2 through sulfate particles is –40 [56]. This means that the environmental impact factor is 1.72E–06*(–40) = –6.9E–05 person-years/kg SO_2.

The uncertainty is the same as for CO_2 (a factor of 2) plus that for the GWP, totally a factor of 2.1 in a log-normal distribution.

6.6.3.7 Impacts on working capacity via climate change

For CO_2, the environmental impact factor for undernutrition was determined in 6.1.3.6 to 4.53E–03 person-hours/kg CO_2. The GWP 100 for SO_2 through sulfate particles is –40 [56]. This means that the environmental impact factor is 4.53E–03*(–40) = –1.81E–01 person-hours/kg SO_2.

The uncertainty is the same as for CO_2 (a factor of 2) plus that for the GWP, totally a factor 2.1 in a log-normal distribution.

6.6.3.8 Impacts on diarrhea diseases via climate change

For CO_2, the environmental impact factor for diarrheal diseases was determined in Sec. 6.1.3.7 to 2.69E–10 person-years/kg CO_2. The GWP 100 for SO_2 through sulfate particles is –40 [56]. This means that the environmental impact factor is 2.69E–10*(–40) = –1.1E–08 person-years/kg SO_2.

The uncertainty is the same as for CO_2 (a factor of 3) plus a factor of 2 for the GWP, which varies significantly between different areas. Totally a factor of $\exp\sqrt{(\ln 3)^2 + (\ln 2)^2} = 3.7$.

6.6.3.9 *Impacts on crop production capacity via climate change*

As described in Sections 6.1.3.8 and 6.1.3.9 there are two pathways by which climate change decreases crop production capacity: (1) changing temperature and precipitation and 2) rising sea level. Here the two impact factors are added. For CO_2, 1.01E–02 + 5.29E–03 = 1.54E–02 kg crop/kg CO_2. The GWP 100 for SO_2 through sulfate particles is –40 [56]. This means that the environmental impact factor is 1.54E–02*(–40) = –6.16E–01 kg crop/kg SO_2.

The uncertainty is the same as for CO_2 (a factor of 3) plus a factor of 2 for the GWP, which varies significantly between different areas, totally $\exp\sqrt{(\ln 3)^2 + (\ln 2)^2} = 3.7$.

6.6.3.10 *Impacts on meat production capacity via climate change*

For CO_2, the environmental impact factor for meat production capacity was determined in Sec. 6.1.3.10 to 3.72E–04 kg meat/kg CO_2. The GWP 100 for SO_2 through sulfate particles is –40 [56]. This means that the environmental impact factor is 3.72E–04*(–40) = –1.49E–02 kg meat/kg SO_2.

The uncertainty is the same as for CO_2 (a factor of 3) plus a factor of 2 for the GWP, which varies significantly between different areas. Totally, a factor 3.7 in a log-normal distribution.

6.6.3.11 *Impacts on fish production capacity via acidification*

An estimation of the impact on fish production capacity can be made from the share of lakes where the critical load for acid deposition are exceeded. Data from Sweden indicates an order of 10% [44] of the lakes although decreasing since 1990s [45]. The acidified lake area in northern Europe and northern America is estimated to be less than 1% of the global lake area subject to fishing. FAO estimated the global freshwater catch to 10 million tons annually. This will correspond to a loss of less than 100 000 tons of fish annually or 1E+08 kg.

As NO_x and NH_3 also cause lake acidification, the contribution from SO_x should not be 100%. The global emission of SO_x is 1.27E+05 Gg (AR5, WGII, Table SM 8.19), which results in 1.27E+05/64*2 = 3969 Gmoles of H +, compared to 2660 Gmoles H + from NO_x, from a NO_x emission of 37.2E+09 kg N, and 2447 Gmole from 4.16E+10 kg NH_3/year. An approximate estimate is that 1 kg SO_x contributes with 1/1.27E+11*3969/(2660 + 3969 + 2447) = 3.44E–12 to fish loss.

The environmental impact factor thus becomes less than 1E+08*3.44E–12 = 3.44E–04 kg fish/kg SO_x.

The uncertainty is large, but it is assumed that the real environmental impact factor is less than the estimate.

6.6.3.12 Impacts on wood production capacity via climate change

For CO_2, the environmental impact factor for wood production capacity was determined in 6.1.3.12 to 0 ± 6.90E–06 m^3 wood/kg CO_2. The GWP 100 for SO_2 through sulfate particles is –40 [56]. This means that the environmental impact factor is 0 ± 2.76E–04 m^3 wood/kg SO_2.

6.6.3.13 Impacts on drinking water production capacity via climate change

The environmental impact factor was determined in 6.1.3.13 to a loss of 9.06E–04 m^3 drinking water production capacity per kg of CO_2. The GWP 100 for SO_2 through sulfate particles is –40 [56]. The environmental impact factor is then 9.06E–04*(–40) = –3.62E–02 m^3 drinking water/kg SO_2.

The uncertainty is the same as for CO_2 (a factor of 3) plus a factor of 2 for the GWP, which varies significantly between different areas. Totally the uncertainty is a factor 3.7 in a log-normal distribution.

6.6.3.14 Impacts on biodiversity via climate change

In 6.1.3.14 the environmental impact factor was estimated to 1.69E–16 shares of the present impact on biodiversity per kg of CO_2. The GWP 100 for SO_2 through sulfate particles is –40 [56]. The environmental impact factor is then 1.69E–16*(–40) = –6.76E–15 shares of the present impact on biodiversity.

The uncertainty is assumed to be a factor of 4 for the CO_2 impact factor and 2 for GWP. The total uncertainty then is a factor of 4.7.

6.6.3.15 Impacts on biodiversity via acidification

According to IUCN database on red listed species, 54 species are threatened by acid rain out of red listed 113,733 species. The anthropogenic emissions of SO_2 for 2010 has been estimated to 100 Tg by Klimont et al. [57].

As NO_x and NH_3 also cause lake acidification, the contribution from SO_x should not be 100%. The total global emission of SO_x is 1.27E+05 Gg

Table 6.15. Summary of environmental impact factors and monetary impact values of SO_x emissions.

Environmental good	Impact indicator	Unit	Pathway	Environmental impact factor	Uncer-tainty	Indicator value ($/unit)	Impact value ($/kgSO$_x$)
human health	YLL	personyears	direct exposure	9.55E–06	2	107,067	1.02E+00
human health	YLL	personyears	secondary particles	5.96E–06	3	107,067	6.38E–01
human health	YLL	personyears	climate change	–5.64E–05	2.7	107,067	–6.04E+00
human health	YLD	personyears	direct exposure	7.50E–06	2	10,707	8.03E–02
human health	YLD	personyears	secondary particles	1.46E–06	2.2	10,707	1.56E–02
human health	undernutrition	personyears	climate change	–6.90E–05	2.1	6,424	–4.43E–01
human health	working capacity	personhours	climate change	–1.81E–01	2.1	30	–5.43E+00
human health	diarrhea	personyears	climate change	–1.10E–08	3.7	11,242	–1.24E–04
crop	production capcity	kg	climate change	–6.16E–01	3.7	0.289	–1.78E–01
meat	production capcity	kg	climate change	–1.49E–02	3.7	2.59	–3.86E–02
fish	production capcity	kg	acidification	1.60E–03	upper limit	2.42	3.87E–03
wood	production capcity	m^3	climate change	0	± 2.76E–04	72	0
drinking water	production capcity	m^3	climate change	–3.62E–02	3.7	1.87	–6.77E–02
biodiversity	share of threat to redlisted species	dimensionless	climate change	–6.76E–15	4.7	7.61E+10	–5.14E–04
biodiversity	share of threat to redlisted species	dimensionless	acidification	2.83E–15	1.5	7.61E+10	2.15E–04
all	CO_2 emission	kg	corrosion	7.68E–03	3	3.76E–01	2.89E–03
						sum	–1.04E+01

(AR5, WGII, Table SM 8.19), which results in 1.27E5/64*2 = 3969 Gmoles of H +, compared to the = 2660 Gmoles H + from NO_x, from a NO_x emission of 37.2E+09 kg N, and 2447 Gmole from 4.16E+10 kg NH_3/year. 1 kg SO_x then contributes with 1/1.27E+11*3969/(2660 + 3969 + 2447) = 3.44E–12 to decreased biodiversity.

The environmental impact indicator is thus 54/113733*3.44E–12 = 1.63E–15 shares of the present impact on biodiversity/kg SO_2.

The uncertainty in the numbers used is moderate. An aggregated uncertainty of a factor of 1.5 is assumed.

6.6.3.16 Indirect impacts on CO_2 emissions via corrosion

The global replacement of steel caused by corrosion due to SO_x is estimated as 3.2E+10 kg/year. The estimation is based on an assumption of a 5% reduction in the usage time for exposed steel constructions and a world steel production of around 1.6 billion tons 2013. Export statistics indicate that around 40% of the steel may be exposed to air although coated (World Steel Association 2014).

When producing 1 kg of steel, there is an emission of around 1.2 kg of CO_2 (including benefits from recycling). This will cause an extra global emission of 32000000000*0.4*0.05*1.2 = 7.86E+08 kg CO_2.

The anthropogenic emissions of SO_2 for 2010 has been estimated as 100 Tg by Klimont et al. [57]. The contribution to the impact is thus 1E–11 per kg SO_2.

The "environmental impact factor" is then 7.68E+08*1E–11 = 7.68E–3 kg CO_2/kg SO_2.

The uncertainty is high in the estimate of how much steel that needs to be replaced. A factor of 3 is assumed for the uncertainty.

6.6.4 Summary and monetary valuation of SO_x emissions

A summary of monetary values of impact on environmental goods from sulfur oxides are shown in Table 6.15. The moderation of climate change from the cooling effect of secondary sulfate aerosols result in negative impact costs, i.e., monetary benefits. Direct negative effects on the respiratory system from exposure of SO_x and sulfate particles is an order of magnitude less costly than the benefits.

References

1. Field, C.B.e.a., *Climate Change 2014, Impacts, Adaption and Vulnerability, Part A Global and Sectorial Aspects*. 2014, WMO, UNEP.

2. Weihe, Urban Climatology and its Applications with Special Regard to Tropical Areas. In *Proceedings of the Technical Conference, Mexico D.F, 26–30 November 1984*, 1986, WMO: Geneva.
3. Oke, Urban Climatology and its Applications with Special Regard to Tropical Areas. In *Proceedings of the technical conference, Mexico D.F, 26–30 November 1984*, 1986, WMO: Geneva.
4. Baccini, M., T. Kosatsky and A. Biggeri, Impact of Summer Heat on Urban Population Mortality in Europe during the 1990s: An Evaluation of Years of Life Lost Adjusted for Harvesting. *PLOS ONE*, 2013. **8**(7).
5. Stocker, T.F. et al., Summary for Policymakers. In*: Climate Change 2013: The Physical Science Basis. Contribution of Working Group I to the Fifth Assessment Report of the Intergovernmental Panel on Climate Change.* 2013, IPCC: Cambridge, United Kingdom and New York, NY, USA.
6. Stocker, T.F. et al., *Climate Change 2013: The Physical Science Basis. Contribution of Working Group I to the Fifth Assessment Report of the Intergovernmental Panel on Climate Change* 2013, IPCC: Cambridge, United Kingdom and New York, NY, USA.
7. Lozano, R. et al., *Global and regional mortality from 235 causes of death for 20 age groups in 1990 and 2010: a systematic analysis for the Global Burden of Disease Study 2010.* The Lancet, 2012. **380**(9859): 2095–2128.
8. Salomon, J.A. et al., Healthy life expectancy for 187 countries, 1990–2010: a systematic analysis for the Global Burden Disease Study 2010. *The Lancet*, 2012. **380**(9859): 19.
9. Lloyd, S., R.S. Kovats and Z. Chalabi, Climate Change, Crop Yields, and Undernutrition: Development of a Model to Quantify the Impact of Climate Scenarios on Child Undernutrition. *Environmental Health Perspectives*, 2011. **119**(12): 7.
10. Kolstad Erik, W. and A. Johansson Kjell, Uncertainties Associated with Quantifying Climate Change Impacts on Human Health: A Case Study for Diarrhea. *Environmental Health Perspectives*, 2011. **119**(3): 299–305.
11. Dunne, J.P., R.J. Stouffer and J.G. John, Reductions in labour capacity from heat stress under climate warming. *Nature Climate Change*, 2013. **3**: 4.
12. Kjellstrom, T.K., R. Sari, Simon J. Lloyd, Holt, Tom, Tol and S.J. Richard. *The direct impact of climate change on regional labour productivity, in Working paper. 2008,* The Economic and Social Research Institute (ESRI): Dublin.
13. Vos, T. et al., Years lived with disability (YLDs) for 1160 sequelae of 289 diseases and injuries 1990–2010: a systematic analysis for the Global Burden of Disease Study 2010. *The Lancet*, 2012. **380**(9859): 2163–2196.
14. *FAOSTAT* 2019, FAO.
15. Narita, D., K. Rehdanz and R.S.J. Tol, Economic costs of ocean acidification: a look into the impacts on global shellfish production. *Climatic Change*, 2012. **113**(3): 1049–1063.
16. Kroeker, K.J. et al., Meta-analysis reveals negative yet variable effects of ocean acidification on marine organisms. *Ecology Letters*, 2010. **13**(11): 1419–1434.
17. Kramer, K., I. Leinonen and D. Loustau, The importance of phenology for the evaluation of impact of climate change on growth of boreal, temperate and Mediterranean forests ecosystems: an overview. *Int. J. Biometeorol.* 2000. **44**: 9.
18. Schewe, J. et al. *Multimodel assessment of water scarcity under climate change.* Proceedings of the National Academy of Sciences, 2014. **111**(9): 6.
19. Weinstock, B., Carbon Monoxide: Residence Time in the Atmosphere. *Science*, 1969. **166**(3902): 224–225.
20. Lelieveld, J. et al. Model calculated global, regional and megacity premature mortality due to air pollution. *Atmos. Chem. Phys.* 2013. **13**(14): 7023–7037.
21. Olivier, J.G.J., et al., A 1990 global emission inventory of anthropogenic sources of carbon monoxide on 1°×1° developed in the framework of EDGAR/GEIA. *Chemosphere - Global Change Science*, 1999. **1**(1): 1–17.

22. Stroppiana, D. et al. Comparison of global inventories of CO emissions from biomass burning derived from remotely sensed data. *Atmos. Chem. Phys.* 2010. **10**(24): 12173–12189.
23. Labouze, E. et al., Photochemical ozone creation potentials. *The International Journal of Life Cycle Assessment*, 2004. **9**(3): 187–195.
24. Tsuruta, A. et al., Global methane emission estimates for 2000–2012 from CarbonTracker Europe-CH4 v1.0. *Geosci. Model Dev.* 2017. **10**(3): 1261–1289.
25. Crippa, M., et al., Gridded emissions of air pollutants for the period 1970–2012 within EDGAR v4.3.2. *Earth System Science Data* 2018. **10**(4): 1987–2013.
26. Stevenson, D.S., et al., Tropospheric ozone changes, radiative forcing and attribution to emissions in the Atmospheric Chemistry and Climate Model Intercomparison Project (ACCMIP). *Atmos. Chem. Phys.* 2013. **13**(6): 3063–3085.
27. Cheng, H.R., et al., Assessing photochemical ozone formation in the Pearl River Delta with a photochemical trajectory model. *Atmospheric Environment*, 2010. **44**(34): 4199–4208.
28. van Dingenen, R. et al., The global impact of ozone on agricultural crop yields under current and future air quality legislation. *Atmospheric Environment*, 2009. **43**: 15.
29. Schlesinger, W.H., *On the fate of anthropogenic nitrogen.* Proceedings of the National Academy of Sciences of the United States of America, 2009. **106**(1): 203–208.
30. Perring, A.E., S.E. Pusede and R.C. Cohen, An Observational Perspective on the Atmospheric Impacts of Alkyl and Multifunctional Nitrates on Ozone and Secondary Organic Aerosol. *Chemical Reviews*, 2013. **113**(8): 5848–5870.
31. Chang, T.Y., J.M. Norbeck and B. Weinstock, An estimate of the nitrogen oxides (NOx) removal rate in an urban atmosphere. *Environmental Science & Technology*, 1979. **13**(12): 1534–1537.
32. Michalski, G., S.K. Bhattacharya and G. Girsch, NO_x cycle and the tropospheric ozone isotope anomaly: an experimental investigation. *Atmos. Chem. Phys.* 2014. **14**(10): 4935–4953.
33. Arnaud, H., At the boundary between anthropogenic and environmental systems: the neglected emissions of indirect nitrous oxide. *The International Journal of Life Cycle Assessment*, 2019. **24**: 412–418.
34. Dockery, D.W., et al., An Association between Air Pollution and Mortality in Six U.S. Cities. *N. Engl. J. Med.* 1993. **329**: 1753–1759.
35. Yang, Y., et al., Short-term and long-term exposures to fine particulate matter constituents and health: A systematic review and meta-analysis. *Environmental Pollution*, 2019. **247**: 874–882.
36. Hill, A.B., The environment and disease: Association or causation? *Journal of the Royal Society of Medicine*, 2015. **108**(1): 32–37.
37. Mukherjee, A. and M. Agrawal, A Global Perspective of Fine Particulate Matter Pollution and Its Health Effects. *In*: *Reviews of Environmental Contamination and Toxicology*, P. de Voogt, Ed., 2018, Springer International Publishing: *Cham*. pp. 5–51.
38. Ji, X. et al., Histone modification in the lung injury and recovery of mice in response to PM2.5 exposure. *Chemosphere*, 2019. **220**: 127–136.
39. Derwent, R.G., Jenkin, M.E., and S.M. Saunders, Photochemical ozone creation potentials for a large number of reactive hydrocarbons under european conditions. *Atmospheric Environment*, 1996. **30**(2): 181–199.
40. Cohen, A.J., et al., *Estimates and 25-year trends of the global burden of disease attributable to ambient air pollution: An analysis of data from the Global Burden of Diseases Study 2015. The Lancet*, 2017. **389**(May 13): 1907–18.
41. Diaz, R.J. and R. Rosenberg, Spreading dead zones and consequences for marine ecosystems. *Science*, 2008. **321**(5891): 926–929.
42. Galloway, J.N., et al., Nitrogen Cycles: Past, Present, and Future. *Biogeochemistry*, 2004. **70**(2): 153–226.

43. Dentener, F., et al., Nitrogen and sulfur deposition on regional and global scales: A multimodel evaluation. *Global Biogeochemical Cycles,* 2006. **20**(4).
44. Bostedt, G., et al., *Acidification Remediation Alternatives: Exploring the Temporal Dimension with Cost Benefit Analysis.* Vol. 39. 2010, pp. 40–48.
45. Garmo, Ø.A., et al., Trends in Surface Water Chemistry in Acidified Areas in Europe and North America from 1990 to 2008. *Water, Air, & Soil Pollution,* 2014. **225**(3): 1880.
46. Reich, P.B., Quantifying plant response to ozone: A unifying theory. *Tree Physiology,* 1987. **3**: 63–91.
47. Cailleret, M., et al., Ozone effects on European forest growth—Towards an integrative approach. *Journal of Ecology,* 2018. **106**(4): 1377–1389.
48. Nadelhoffer, K.J., et al., Nitrogen deposition makes a minor contribution to carbon sequestration in temperate forests. *Nature,* 1999. **398**: 145.
49. *IUCN Redlist.* 2014.
50. Seppälä, J., et al., Country-Dependent Characterisation Factors for Acidification and Terrestrial Eutrophication Based on Accumulated Exceedance as an Impact Category Indicator. *Int. J. LCA,* 2006. **11**(6): 403–416.
51. Allen, M.R., et al., New use of global warming potentials to compare cumulative and short-lived climate pollutants. *Nature Climate Change,* 2016. **6**(8): 773–776.
52. Sun, L., et al., Combined effects of nitrogen deposition and biochar application on emissions of N2O, CO2 and NH3 from agricultural and forest soils. *Soil Science and Plant Nutrition,* 2014. **60**(2): 254–265.
53. Suntharalingam, P., et al., Quantifying the impact of anthropogenic nitrogen deposition on oceanic nitrous oxide. *Geophysical Research Letters,* 2012. **39**(7).
54. Consulting, I., *Human Health Benefits of Stratospheric Ozone Protection,* O.f.o.A.a.R. Global Programs Division, U.S. Environmental Protection Agency, Editor. 2006: Washington, DC 20460.
55. Ravishankara, A.R., J.S. Daniel and R.W. Portmann, Nitrous Oxide (N<sub>2</ sub>O): The Dominant Ozone-Depleting Substance Emitted in the 21st Century. *Science,* 2009. **326**(5949): 123–125.
56. Fuglestvedt, J.S., et al., Transport impacts on atmosphere and climate: Metrics. *Atmospheric Environment,* 2010. **44**(37): 4648–4677.
57. Klimont, Z., S.S. J and J. Cofala, The last decade of global anthropogenic sulfur dioxide: 2000–2011 emissions. *Environmenal Research Letters,* 2013. **8**: 6.
58. Rasch, P.J., et al., A description of the global sulfur cycle and its controlling processes in the National Center for Atmospheric Research Community Climate Model, Version 3. *Journal of Geophysical Research: Atmospheres,* 2000. **105**(D1): 1367–1385.
59. Lee, C.-L. and P. Brimblecombe, Anthropogenic contributions to global carbonyl sulfide, carbon disulfide and organosulfides fluxes. *Earth-Science Reviews,* 2016. **160**: 1–18.
60. Stedman, J.R., E. Linehan and K. King, *Quantification of the Health Effects of Air Pollution in the UK for the Review of the National Air Quality Strategy.* 1999, AEA Technology, National Environmental Technology Centre: Abingdon.
61. Guerreiro, C.B.B., V. Foltescu and F. de Leeuw, Air quality status and trends in Europe. *Atmospheric Environment,* 2014. **98**: 376–384.
62. Li, C., et al., *Author Correction:* India is overtaking China as the world's largest emitter of anthropogenic sulfur dioxide (Scientific Reports (2017) DOI: 10.1038/s41598-017-14639-8). *Scientific Reports,* 2018. **8**(1).
63. Crutzen, P.J., Albedo Enhancement by Stratospheric Sulfur Injections: A Contribution to Resolve a Policy Dilemma? *Climatic Change,* 2006. **77**(3): 211.
64. Reiss, R., et al., Evidence of health impacts of sulfate-and nitrate-containing particles in ambient air. *Inhal Toxicol.,* 2007. **19**(5): 419–449.

7

Emissions of Particles to Air

Particles in air, aerosols, are much more difficult to classify than gases. Particles differ in size, form and composition. The reason it still is meaningful to classify them and link them to specific impacts is that most particles in ambient air originate from a limited number of sources and processes.

There are two different formation processes: disintegration/dispersion and condensation. Dispersion aerosols are typically larger than 10 μm aerodynamical diameter, while condensation aerosols are around 1 μm and less. Dispersion aerosols have a short residence time in air: minutes to hours. Dispersion aerosols therefore often have a local character. Condensation aerosols have residence times of days to weeks and form a mixture of particles within a larger area, like an urban area. The energy needed to create a particle is high, and the smaller the particle, the more energy is needed per mass unit. In a condensation process the energy available is high. Fumes from high temperature processes like steel ovens and combustion processes contain very small particles, below 0.1 μm. Condensation of gases with high volatility gives off larger particles. Water droplets in clouds may be around 50 μm. Small particles below 0.1 μm have high diffusion velocities and form aggregates relatively fast. In air, 0.3 μm particles are the most stable ones. They have low diffusion velocity and a low settling velocity. A typical size distribution for particles in ambient air is shown in Figure 7.1.

As the condensation aerosol becomes older the peak diameter moves to the left and approaches 0.3 μm. The dispersion aerosol varies quite a lot depending on the surface dryness and wind speed.

The minima between the peaks is often around 2–3 μm. That is why it is logical to have a particle size less than 2.5 μm. Another reason is that

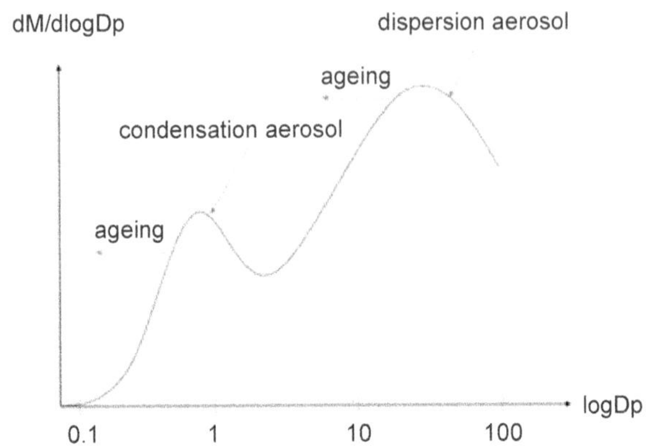

Figure 7.1. Typical size distribution in ambient air. M is the mass of particles with the aerodynamic diameter D_p. The peak particle size decreases as the aerosol ages, and large particles fall out.

particles less than 2.5 μm aerodynamic diameter penetrate into the alveoli of the lung. Particles larger than 10 μm aerodynamic diameter do not normally enter the respiratory system. They are separated from the air entering the mouth and nose by inertia forces.

There are no measurement methods with a sharp cut-off at 2.5 or 10 μm. Particles below 2.5 μm aerodynamic diameter are in practice particles that penetrate a separator to 50%. This means that samples of PM2.5 particles also contain larger particles. To some extent, this is not a big problem as particle deposition is the respiratory organs also lack a cut-off at a distinct particle size. But there are other factors that complicate the categorization of PM2.5 with respect to deposition in lungs and health effects: The aerodynamic diameter of a certain particle is the diameter of a sphere with the density 1 g/cm^3 that has the same settling velocity in air as the particle. Both the shape of the particle and the hygroscopicity influence the deposition. Fibers may have low deposition velocity in air but may still be easily caught in fine lung structures. In lungs, where the humidity is high, small particles may turn into droplets in milliseconds.

Besides the uncertainties regarding particle categorization in ambient air, there is also measurement uncertainties at sources. Particle fractions are very much defined by its sampling and measurement method. Emission measurements are often different than measurements in ambient air. In case of stacks, there are uncertainties due to condensable material. Measurements are often made at elevated temperatures (e.g., 150°C) to avoid water condensing in the filters. But that may result in that volatile substances like SO_3, PAH, VOC and some metal chlorides, which are in the gas phase in the flue gas, are not included in the particle emissions measure although they will condense on the particles later in the stack plume and be found on particles in ambient air. Sometimes this will cause large confusions, especially when efficient dust removing equipment operating at elevated temperatures is used. Then solid particles are removed, but volatile substances remain [1, 2]. What is called "secondary aerosols" in air may be partly formed through atmospheric reactions and partly from condensation in the stack plume. When assessing environmental impacts of particle emissions, it may not always be clear if condensable material was included or not.

A particular problem is to relate "soot" measures of particle emissions to "Black Carbon (BC)" measures of ambient air aerosols. In both cases it is the light reflection of particles collected on a filter that is determined. But light reflection depends on particle size and composition, and it is difficult to model the relation between emission and concentration in ambient air, as particle size and composition changes in a complex way. A way around this problem is to use the measure "Elemental Carbon, (EC)" as an approximate measure for "soot" and BC. EC and OC, organic carbon, is determined from the CO_2 formed after standardized heating and combustion sequences and is thereby chemically defined components of particles.

7.1 PM2.5

Here PM2.5 is defined as what is passing a particle size separator, e.g., a cyclone, impactor or virtual impactor to 50% at 2.5 μm. Typically, only 10% of particles around of 3 μm pass the separator and 90% of the 2 μm particles.

7.1.1 Specification of the emission

The emission valued is PM2.5 including condensable matter, i.e., PM2.5 particles such as they occur in ambient air, close to the source. Sources of such primary PM2.5 are primarily combustion processes and other high temperature processes. Road dust, crushing and grinding and agriculture chiefly generate coarse particles, but a small part of these may be included in PM2.5.

The emissions valued here are primary PM2.5 emitted anywhere in the world 2015.

7.1.2 Impact pathways

The major environmental impacts of economic significance are health effects from direct exposure to PM2.5 and climate impacts. PM2.5 also influences visibility, but this is not considered to be of significant sustainability concern. Table 7.1 shows the impacts of PM2.5 modeled.

Table 7.1. Pathways by which PM2.5 particles impact on environmental goods.

Environmental good	Impact indicator	Unit	Pathway
human health	YLL	personyears	direct exposure
human health	YLL	personyears	climate change
human health	YLD	personyears	direct exposure
human health	undernutrition	personyears	climate change
human health	working capacity	personhours	climate change
human health	diarrhea	personyears	climate change
crop	production capacity	kg	climate change
meat	production capacity	kg	climate change
fish	production capacity	kg	climate change
wood	production capacity	m^3	climate change
drinking water	production capacity	m^3	climate change
biodiversity	share of threat to redlisted species	dimensionless	climate change

7.1.3 Environmental impact factors

7.1.3.1 YLL via direct exposure

Cohen et al. [3] found that 99.2 million YLL (87.7–111 million with 95% probability) was caused by PM2.5 2015. WHO reports that 85 million YLL is caused by PM2.5 2012.

Global primary emission of PM2.5 was 38 Tg 2005 according to emission inventories reported by EU JRC [4]. Primary emissions of PM2.5 account for about 40% to 80% of PM2.5 in ambient air. Examples of source contributions of PM2.5 are shown in Table 7.2. A best estimate of 60% will be used for the modelling.

Table 7.2. Characterization of the composition of PM2.5 with respect to contributing sources. Anthropogenic and crustal are primary sources.

		Contribution to PM2.5 mass, %				
Place	**Year**	**Anthropogenic**	**Crustal**	**Secondary**	**Primary**	**Reference**
Beijing, China	2001	59	20	19	79	[5]
New Taipei City, Taiwan	2001	56	21	22	77	[6]
Milan, Italy	2001	42	15	43	57	[7]
Barcelona, Spain	1999–2000	49	5	46	54	[8]
New York, USA	2015	38	6	56	44	[9]

The environmental impact factor is thus 8.5E+7*0.6/3.8E+10 = 1.34E–03 YLL/kg PM2.5.

The environmental impact factor will vary significantly depending on local conditions, such as population density and ventilation premises. This introduces an uncertainty in the order of a factor of 3. There is also an uncertainty in the specification of PM2.5 and consequently its health effects. It is probably the black carbon and organic carbon, that contributes most to the health effects while sulfate, nitrates and soil dust have less impact [10]. Considering the variations seen in Table 7.2 this would introduce an uncertainty of a factor of 1.5. The total uncertainty may therefore be represented by a log-normal distribution with a standard deviation corresponding to a factor of exp $\sqrt{(\ln 3)^2 + (\ln 1.5)^2} = 3.2$.

7.1.3.2 YLL via climate change

No GWP values for PM2.5 have been found in IPCC AR5 reports or in the literature. However, GWPs are given for black carbon particles (BC) and organic carbon particles (OC). BC has a GWP100 of 658.6 and OC has a GWP100 of –66.4. The GWP for sulfate particles are given by Fuglestvedt et al. [11] to –40. As most of the anthropogenic emissions of PM2.5 are from burning of fossil fuels, the black carbon GWP will determine the GWP for anthropogenic PM2.5. A best estimate of GWP for PM2.5 emissions of 330 is estimated, assuming that about half of the PM2.5 emissions are BC.

YLL caused by CO_2 was 1.88E–06/kg CO_2 (Table 6.3). The environmental impact factor for PM2.5 is then 1.88E–06*330 = 6.20E–04 YLL/kg PM2.5.

Undernutrition is the dominating environmental impact factor amongst CO_2's impacts on YLL. The uncertainty for that impact factor is estimated to a factor of 2. For PM2.5 there is an additional uncertainty in the GWP100 value of a factor of 2, due to variation between regions. The total uncertainty may therefore be represented by a log-normal distribution with a standard deviation corresponding to a factor of exp $\sqrt{(\ln 2)^2 + (\ln 2)^2} = 2.7$.

7.1.3.3 YLD via direct exposure

Cohen et al. [3] found that 3.9 million (2·6 million to 5·2 million) years lived with disability (YLD) was caused by PM2.5 2015.

Global emission of PM2.5 was 38 Tg 2005 according to EU JRC [4]. Primary emissions of PM2.5 account for about 40 to 80% of PM2.5 in ambient air. Examples of source contributions of PM2.5 are shown in Table 7.2. A best estimate of 60% will be used for the modelling.

The environmental impact factor is thus 3.9E+06*0.6/3.8E+10 = 6.16E–05 YLD/kg PM2.5.

The environmental impact factor will vary significantly depending on local conditions, such as population density and ventilation premises. This introduces an uncertainty in the order of a factor of 3. There is also an uncertainty in the specification of PM2.5 and consequently its health effects. It is probably the black carbon and organic carbon, that contributes most to the health effects while sulfate, nitrates and soil dust have less impact [10]. Considering the variations seen in Table 7.2 this would introduce an uncertainty of a factor of 1.5. The total uncertainty may therefore be represented by a log-normal distribution with a standard deviation corresponding to a factor of exp $\sqrt{(\ln 3)^2 + (\ln 1.5)^2} = 3.2$.

7.1.3.4 Undernutrition via climate change

No GWP values for PM2.5 have been found in IPCC AR5 reports or in the literature. However, GWPs are given for black carbon particles (BC) and organic carbon particles (OC). BC has a GWP100 of 658.6 and OC has a GWP100 of –66.4. The GWP for sulfate particles are given by Fuglestvedt et al. [11] to –40. As most of the anthropogenic emissions are from fossil fuel combustion, the black carbon GWP will determine the GWP for PM2.5. A best estimate of GWP for PM2.5 emissions of 330 is estimated, assuming that about half of the PM2.5 emissions are BC.

Undernutrition caused by CO_2 was 1.72E–06 personyears/kg CO_2 (Table 6.3). The environmental impact factor for PM2.5 is then 1.72E–06*330 = 5.68E–04 personyears/kg PM2.5.

The uncertainty for the impact factor of CO_2's impact on undernutrition was estimated to a factor of 4. For PM2.5 there is an additional uncertainty in the GWP100 value of a factor of 2, due to variation between regions. The total uncertainty may therefore be represented by a log-normal distribution with a standard deviation corresponding to a factor of exp $\sqrt{(\ln 4)^2 + (\ln 2)^2} = 4.7$.

7.1.3.5 Decreased working capacity via climate change

No GWP values for PM2.5 have been found in IPCC AR5 reports or in the literature. However, GWPs are given for black carbon particles (BC) and organic carbon particle (OC). BC has a GWP100 of 658.6 and OC has a GWP100 of –66.4. The GWP for sulfate particles are given by Fuglestvedt et al. [11] to –40. As most of the anthropogenic emissions are from fossil fuel combustion, the black carbon GWP will determine the GWP for PM2.5. A best estimate of GWP for PM2.5 emissions of 330 is chosen for the model, assuming that about half of the PM2.5 emissions are BC.

Loss of working capacity caused by CO_2 is 4.53E–03 personhours/kg CO_2 (Table 6.3). The environmental impact factor for PM2.5 is then 4.53E–03*330 = 1.49 personhours/kg PM2.5.

The uncertainty for the impact factor of CO_2's impact on working capacity was estimated to a factor of 2. For PM2.5 there is an additional uncertainty in the GWP100 value of a factor of 2, due to variation between regions. The total uncertainty may therefore be represented by a log-normal distribution with a standard deviation corresponding to a factor of exp $\sqrt{(\ln 2)^2 + (\ln 2)^2} = 2.7$.

7.1.3.6 Diarrhea via climate change

No GWP values for PM2.5 have been found in IPCC AR5 reports or in the literature. However, GWPs are given for black carbon particles (BC) and organic carbon particles (OC). BC has a GWP100 of 658.6 and OC has a GWP100 of –66.4. The GWP for sulfate particles are given by Fuglestvedt et al. [11] to –40. As most of the anthropogenic emissions are from fossil fuel combustion, the black carbon GWP will determine the GWP for PM2.5. A best estimate of GWP for PM2.5 emissions of 330 is chosen for the model, assuming that about half of the PM2.5 emissions are BC.

CO_2 causes 2.69E–10 personyears/kg CO_2 of diarrhea (Table 6.3). The environmental impact factor for PM2.5 is then 2.6953E–10*330 = 8.89E–08 personyears/kg PM2.5.

The uncertainty for the impact factor of CO_2's impact on diarrhea was estimated to a factor of 3. For PM2.5 there is an additional uncertainty in the GWP100 value of a factor of 2, due to variation between regions. The total uncertainty may therefore be represented by a log-normal distribution with a standard deviation corresponding to a factor of exp $\sqrt{(\ln 3)^2 + (\ln 2)^2} = 3.7$.

7.1.3.7 Crop production capacity via climate change

No GWP values for PM2.5 have been found in IPCC AR5 reports or in the literature. However, GWPs are given for black carbon particles (BC) and organic carbon particle (OC). BC has a GWP100 of 658.6 and OC has a GWP100 of –66.4. The GWP for sulfate particles are given by Fuglestvedt et al. [11] to –40. As most of the anthropogenic emissions are from fossil fuel combustion, the black carbon GWP will determine the GWP for PM2.5. A best estimate of GWP for PM2.5 emissions of 330 is chosen for the model, assuming that about half of the PM2.5 emissions are BC.

1 kg CO_2 causes 1.42E–02 kg decreased production capacity (Table 6.3). The environmental impact factor for PM2.5 is thus 1.42E–02*330 = 4.69 kg crop/kg PM2.5.

The uncertainty for the impact factor of CO_2's impact on crop production capacity was estimated to a factor of 3. For PM2.5 there is an additional uncertainty in the GWP100 value of a factor of 2, due to variation between regions. The total uncertainty may therefore be represented by a log-normal distribution with a standard deviation corresponding to a factor of $\exp\sqrt{(\ln 3)^2 + (\ln 2)^2} = 3.7$.

7.1.3.8 Meat production capacity via climate change

No GWP values for PM2.5 have been found in IPCC AR5 reports or in the literature. However, GWPs are given for black carbon particles (BC) and organic carbon particle (OC). BC has a GWP100 of 658.6 and OC has a GWP100 of –66.4. The GWP for sulfate particles are given by Fuglestvedt et al. [11] to –40. As most of the anthropogenic emissions are from fossil fuel combustion, the black carbon GWP will determine the GWP for PM2.5. A best estimate of GWP for PM2.5 emissions of 330 is chosen for the model, assuming that about half of the PM2.5 emissions are BC.

1 kg CO_2 causes 3.72E–04 kg decreased meat production capacity (Table 6.3). The environmental impact factor for PM2.5 is thus 3.72E–04*330 = 1.23E–01 kg meat/kg PM2.5.

The uncertainty for the impact factor of CO_2's impact on meat production capacity was estimated to a factor of 3. For PM2.5 there is an additional uncertainty in the GWP100 value of a factor of 2, due to variation between regions. The total uncertainty may therefore be represented by a log-normal distribution with a standard deviation corresponding to a factor of $\exp\sqrt{(\ln 3)^2 + (\ln 2)^2} = 3.7$.

7.1.3.9 Fish production capacity via climate change

No GWP values for PM2.5 have been found in IPCC AR5 reports or in the literature. However, GWPs are given for black carbon particles (BC) and organic carbon particle (OC). BC has a GWP100 of 658.6 and OC has a GWP100 of –66.4. The GWP for sulfate particles are given by Fuglestvedt et al. [11] to –40. As most of the anthropogenic emissions are from fossil fuel combustion, the black carbon GWP will determine the GWP for PM2.5. A best estimate of GWP for PM2.5 emissions of 330 is chosen for the model, assuming that about half of the PM2.5 emissions are BC.

1 kg CO_2 causes 2.92E–05 kg decreased fish production capacity (Table 6.3). The environmental impact factor for PM2.5 is thus 2.92E–05*330 = 9.64E–03 kg fish/kg PM2.5.

The uncertainty for the impact factor of CO_2's impact on fish production capacity was estimated to a factor of 2. For PM2.5 there is an additional uncertainty in the GWP100 value of a factor of 2, due to variation between regions. The total uncertainty may therefore be represented by a log-normal distribution with a standard deviation corresponding to a factor of exp $\sqrt{(\ln 2)^2 + (\ln 2)^2} = 2.7$.

7.1.3.10 Wood production capacity via climate change

No GWP values for PM2.5 have been found in IPCC AR5 reports or in the literature. However, GWPs are given for black carbon particles (BC) and organic carbon particle (OC). BC has a GWP100 of 658.6 and OC has a GWP100 of –66.4. The GWP for sulfate particles are given by Fuglestvedt et al. [11] to –40. As most of the anthropogenic emissions are from fossil fuel combustion, the black carbon GWP will determine the GWP for PM2.5. A best estimate of GWP for PM2.5 emissions of 330 is chosen for the model, assuming that about half of the PM2.5 emissions are BC.

I kg CO_2 causes 0 m^3 decreased wood production capacity (Table 6.3). The environmental impact factor for PM2.5 is thus 0*330 = 0 kg wood/kg PM2.5.

The uncertainty for the impact factor of CO_2's impact on wood production capacity was estimated to 6.90E–06 m^3/kg CO_2. This corresponds to 2.28E–03 m^3 wood per kg PM2.5. For PM2.5 there is an additional uncertainty in the GWP100 value of a factor of 2, due to variation between regions.

7.1.3.11 Drinking water production capacity via climate change

No GWP values for PM2.5 have been found in IPCC AR5 reports or in the literature. However, GWPs are given for black carbon particles (BC) and organic carbon particle (OC). BC has a GWP100 of 658.6 and OC has a GWP100 of –66.4. The GWP for sulfate particles are given by Fuglestvedt et al. [11] to –40. As most of the anthropogenic emissions are from fossil fuel combustion, the black carbon GWP will determine the GWP for PM2.5. A best estimate of GWP for PM2.5 emissions of 330 is chosen for the model, assuming that about half of the PM2.5 emissions are BC.

1 kg CO_2 causes 9.06E–04 m^3 decreased drinking water production capacity (Table 6.3). The environmental impact factor for PM2.5 is thus 9.06E–04*330 = 0.299 m^3 drinking water/kg PM2.5.

The uncertainty for the impact factor of CO_2's impact on drinking water production capacity was estimated to a factor of 3. For PM2.5 there is an additional uncertainty in the GWP100 value of a factor of 2, due to variation between regions. The total uncertainty may therefore be represented by a log-normal distribution with a standard deviation corresponding to a factor of exp $\sqrt{(\ln 3)^2 + (\ln 2)^2} = 3.7$.

7.1.3.12 Decrease of biodiversity via climate change

No GWP values for PM2.5 have been found in IPCC AR5 reports or in the literature. However, GWPs are given for black carbon particles (BC) and organic carbon particle (OC). BC has a GWP100 of 658.6 and OC has a GWP100 of –66.4. The GWP for sulfate particles are given by Fuglestvedt et al. [11] to –40. As most of the anthropogenic emissions are from fossil fuel combustion, the black carbon GWP will determine the GWP for PM2.5. A best estimate of GWP for PM2.5 emissions of 330 is chosen for the model, assuming that about half of the PM2.5 emissions are BC.

1 kg CO_2 causes 1.69E–16 shares of threat to red-listed species (Table 6.3). The environmental impact factor for PM2.5 is thus 1.69E–16*330 = 5.58E–14 shares of threat to red-listed species/kg PM2.5.

The uncertainty for the impact factor of CO_2's impact on biodiversity was estimated to a factor of 4. For PM2.5 there is an additional uncertainty in the GWP100 value of a factor of 2, due to variation between regions. The total uncertainty may therefore be represented by a log-normal distribution with a standard deviation corresponding to a factor of exp $\sqrt{(\ln 4)^2 + (\ln 2)^2} = 4.7$.

7.1.4 Summary and monetary valuation of PM2.5 emissions

A summary of environmental impact factors and monetary impact values are shown in Table 7.3.

The monetary value of the environmental impacts of PM2.5 is startlingly high, 268 $/kg. The largest contribution to the monetary value of the environmental impact of PM2.5 is YLL via direct exposure, but the contributions from impacts on YLL and working capacity caused by climate change is also significant.

Table 7.3. Summary of environmental impact factors and monetary impact values of emission of 1 kg PM2.5.

Impact indicator	Unit	Pathway	Environmental impact factor	Uncertainty	Indicator value ($/unit)	Impact value ($/kg PM2.5)
YLL	personyears	direct exposure	1.34E–03	3.2	107,067	1.44E+02
YLL	personyears	climate change	6.20E–04	2.7	107,067	6.64E+01
YLD	personyears	direct exposure	6.16E–05	3.2	107,067	6.59
undernutrition	personyears	climate change	5.68E–04	4.7	6,424	3.65
working capacity	personhours	climate change	1.49E+00	2.7	30	4.48E+01
diarrhea	personyears	climate change	8.89E–08	3.7	11,242	1.00E–03
production capacity	kg	climate change	4.69E+00	3.7	0.289	1.35
production capacity	kg	climate change	1.23E–01	3.7	2.59	3.18E–01
production capacity	kg	climate change	0.009636	2.7	2.42	2.97E–01
production capacity	m^3	climate change	2.28E–03	0.0	72	1.64E–01
production capacity	m^3	climate change	2.99E–01	3.7	1.87	5.59E–01
share of threat to redlisted species	dimensionless	climate change	5.58E–14	4.7	7.61E+10	4.24E–03
					sum	2.68E+02

7.2 Polycyclic Aromatic Hydrocarbons, PAH

Polycyclic aromatic hydrocarbons, PAH is a subordinate to PAC, polycyclic aromatic compounds, which also contains substituted polycyclic aromatic compounds, such as dinitro pyrenes. PAH is a part of PM2.5 but may also be part of larger particles and are present in gas phase. Normally PAHs is generated during incomplete combustion. PAH also exist in asphalt, and road dust. Much of the knowledge of PAHs carcinogenicity to humans comes from studies in work environments, where there is a mixture of PAHs [12]. Single PAHs are studied in animal tests.

7.2.1 Specification of the emission

In practice emissions of PAHs are defined by the measurement method. When sampling on filters and gas phase absorbents and using gas chromatography and extraction by acetone or other solvents, about 40 PAC substances (Table 7.4) may be identified.

The preparation of the samples for the GC method is quite time consuming and expensive. If high-pressure liquid chromatography (HPLC) is used the preparation step is quicker and more samples can be analyzed on the same budget. However, by HPLC it is only possible to analyze about ten PAHs (Table 7.5). Even if these PAHs are those that occur in highest concentrations, PAH-values determined by the HPLC method tend to be lower than the GC values.

The emissions valued here are PAHs emitted anywhere in the world 2015 and include the substances in Table 7.4.

As normal amounts of PAH in PM2.5 are already accounted for in the valuation of PM2.5 impacts, care must be taken in order to avoid double counting. This may be done through only using the monetary values for PAH in emissions from sources known to have high PAH concentrations like coke ovens, Al smelters, biomass burning and diesel exhausts.

7.2.2 Impact pathways

Some PAHs are proven carcinogenic to humans, e.g., benzo(a)pyrene, some are classified as probably or possibly carcinogenic. PAHs cause lung cancer after inhalation.

7.2.3 Environmental impact factors

7.2.3.1 YLL from direct exposure

Globally, incremental lifetime lung cancer risk (ILCR) induced by ambient PAH exposure is 3.1E−5 [13]. Mortality in lung cancer is about 80% [14].

Table 7.4. PAC compounds typically detected by GC methods.

Substance name	Substance name
Acenaphtylene	Benzo(b)fluorene
Acenaphtene	1-Metylpyrene
4-Metylbiphenyl	Benz(a)anthracene
Dibenzofuran	Chrysene
Fluorene	Triphenylene
9-Metylfluorene	Naphtancene
9,10-Dihydroanthracene	Benzo(bjk)fluoranthenes
1-Metylfluorene	Benzo(a)pyrene
Dibenzothiophene	Benzo(e)pyrene
Phenanthrene	Perylene
Anthracene	3-Methylcholanthrene
2-Metylanthracene	m-Quaterphenyl
1-Metylanthracene	Indeno(1,2,3-cd)pyrene
9-Metylanthracene	Dibenz(a,h)anthracene
3,6-Dimethylphenantrene	Picene
1,2-Dihydropyrene	1,2,3,4-Dibenzanthracene
Fluoranthene	Benzo(g,h,i)perylene
Pyrene	Anthanthrene
Benzo(a)fluorene	

Table 7.5. PAC compounds typically detected by HPLC methods.

Substance name
Phenanthrene
Anthracene
Flouranthene
Pyrene
Benzo(a)anthracene
Chrysene
Benzo(b)fluoranthene
Benzo(k)fluoranthene
Benzo(a)pyrene
Benzo(ghi)perylene
Indeno(cd)pyrene

The YLL per year is then 0.8*0.000031/72*7200000000*10 = 2.48E+04, where 72 is the average lifetime, 7200000000 is the world population in 2015 and 10 is the average YLL per fatal case. The average YLL per fatal lung cancer is estimated from studies in Norway and Iran [15, 16]. Iran is assumed to represent the developing world and Norway the developed world. Yanxu Zhang et al. estimated the global PAH16 emission to be 520 Gg/yr [17]. The environmental impact factor thus becomes 2.48E+04/5.20E+08 = 4.77E–05 YLL/kg PAH.

Even if the cancer risk for benzo(a)pyrene is well known, quantitative knowledge of the risk by other PAHs, which are classified as possible or probable carcinogens, is lacking. An uncertainty of a factor of 3 is assumed.

7.2.3.2 Personyears of cancer from direct exposure

Globally, incremental lifetime lung cancer risk (ILCR) induced by ambient PAH exposure is 3.1E–5 [13]. The YLL per year is then 0.000031/72*7200000000*1.56 = 4.81E+3, where 72 is the average lifetime, 7200000000 is the world population 2015 and 1.56 is the average duration of the disease per case. The value 1.56 is derived from the total global YLD estimate (355000/year 2010) from Vos et al., from the DALY factor for lung cancer (0.15) [18], and from the global mortality in lung cancers 1521000/year [19]. 355000/(0.15*1521000) = 1.56. [20] Yanxu Zhang et al. estimated the global PAH16 emission to 520 Gg/yr [17]. The environmental impact factor thus becomes 4.81E+03/5.20E+08 = 9.25E–06 personyears/kg PAH.

Even if the cancer risk for benzo(a)pyrene is well known, the quantitative knowledge of PAHs that are classified as possible or probable carcinogens is lacking. An uncertainty of a factor of 3 is assumed.

7.2.4 Summary and monetary valuation of PAH emissions

A summary of environmental impact factors and monetary impact values are shown in Table 7.6.

It is striking that the monetary value of PAH impacts is much lower than that of PM2.5. But the PAH that is modeled here are both gas phase and particle-bound PAH. The largest part of PAH is in the gas phase. The most potent PAH, benzo(a)pyrene is only in the particle phase and accounts globally for between 0.05% to 2.08% of the PAH emissions.

Table 7.6. Summary of environmental impact factors and monetary impact values of emission of 1 kg PAH.

Environmental good	Impact indicator	Unit	Pathway	Environmental impact factor	Uncertainty	Indicator value ($/unit)	Impact value ($/kg PAH)
human health	YLL	personyears	direct exposure	4.77E–05	3.2	107,067	5.11E+00
human health	cancer	personyears	direct exposure	9.25E–06	3.2	21,413	1.98E–01
						sum	5.31E+00

7.3 Arsenic

7.3.1 Specification of the emission

As is emitted as part of particles, but gaseous forms exist, even in ambient air. It introduces uncertainty in what data actually represents.

The emission valued is arsenic in both gaseous and particle forms emitted into the air anywhere in the world in 2015.

7.3.2 Impact pathways

Inhaled As cause cancer and increase mortality in cardiovascular diseases.

7.3.3 Environmental impact factors

7.3.3.1 YLL via cancer

USEPA estimates the lifetime risk for cancer to be 4.3E–3/μg/m^3 of As in air. Arsenic causes cancer in lung, bladder and skin. The mortality for lung cancer in the is 83% (National Cancer Institute, http://seer.cancer.gov). The average YLL per fatal lung cancer is estimated to be 10 years from studies in Norway and Iran [15, 16]. Iran is assumed to represent the developing world and Norway the developed world.

The fate of an As particle is assumed to be the same as an average PM10 particle. The global emission of PM10 is estimated to be 63 Tg/yr and the global average PM10 concentration to 61 μg/m^3. Primary emissions contribute by 40% to the total particle mass. This means that 1 kg of As emitted to air will result in a concentration of 0.4*61/63E+09 = 3.78E–10 μg/m^3 and an impact of 4.3E–3*0.83*10*3.78E–10/70*7.2E+09= 1.39E–03 YLL per year.

The uncertainty is assumed to be a factor of 3, much due to the simple fate model.

7.3.3.2 YLL via cardiovascular diseases

Lifetime risk is 0.01 per ug/m^3 [21]. The marginal reduction of life expectancy is estimated to be 5.0 years based on WHO estimates of global YLL/years for different age groups and an average life expectancy of 70 years.

The fate of an As particle is assumed to be the same as an average PM10 particle. The global emission of PM10 is estimated to be 63 Tg/yr and the global average PM10 concentration to 61 μg/m^3. Primary emissions contribute by 40% to the total particle mass. This means that 1 kg of As emitted to air will result in a concentration of 0.4*61/63E+09 = 3.78E–10 μg/m^3 and an impact of 0.01*5*3.78E–10/70*7.2E+09= 1.94E–03 YLL per year.

The uncertainty is assumed to be a factor of 3, much due to the simple fate model.

7.3.3.3 Personyear of disability through cancer

USEPA estimates the lifetime risk for cancer to be 4.3E–3/μg/m^3 of As in air. The average morbidity time for cancers in lung, bladder and skin is assumed to be three years.

The fate of an As particle is assumed to be the same as an average PM10 particle. The global emission of PM10 is estimated to be 63 Tg/yr and the global average PM10 concentration to 61 μg/m^3. Primary emissions contribute by 40% to the total particle mass. This means that 1 kg of As emitted to air will result in a concentration of 0.4*61/63E+09 = 3.78E–10 μg/m^3 and an impact of 4.3E–3*3*3.78E–10/70*7.2E+09 = 5.02E–04 personyears per year.

The uncertainty is assumed to be a factor of 3, much due to the simple fate model.

7.3.4 Summary and monetary valuation

A summary of environmental impact factors and monetary impact values is shown in Table 7.7.

It is striking that the monetary impact value is only slightly higher than that for PM2.5. Maybe it has something to do with how risk estimates for As and PM2.5 are done. Risk estimates for As relies on work environment observations, while the risk for PM2.5 is determined in cohort studies on large population groups. They are risks for different groups under different exposure situations.

7.4 Cadmium

7.4.1 Specification of the emission

The emission valued is Cadmium emitted to air anywhere in the world in 2015.

7.4.2 Impact pathways

Cadmium causes several types of cancers when inhaled.

Table 7.7. Summary of environmental impact factors and monetary impact values of emission of 1 kg As.

Environmental good	Impact indicator	Unit	Pathway	Environmental impact factor	Uncertainty	Indicator value ($/unit)	Impact value ($/kg As)
human health	YLL	personyears	cancer	1.39E–03	3	107,067	1.49E+02
human health	YLL	personyears	CV diseases	1.94E–03	3	107,067	2.08E+02
human health	cancer	personyears	direct exposure	1.79E–05	3	21,413	3.83E–01
						sum	3.57E+02

7.4.3 Environmental impact factors

7.4.3.1 YLL from cancer

USEPA estimates the lifetime risk for cancer to be 1.8E–3/μg/m^3 of Cd in air. The mortality for all sorts of cancer in the US was 34% in 2011. (National Cancer Institute, http://seer.cancer.gov). The average YLL per fatal lung cancer is estimated to 10 years from studies in Norway and Iran [15, 16]. Iran is assumed to represent the developing world and Norway the developed world. The global average life expectancy is 70 years.

The fate of a Cd particle is assumed to be the same as an average PM10 particle. The global emission of PM10 is estimated to be 63 Tg/yr and the global average PM10 concentration to 61 μg/m^3. Primary emissions contribute by 40% to the total particle mass. This means that 1 kg of Cd emitted to air will result in a concentration of 0.4*61/63E+09 = 3.78E–10 μg/m^{3}and an impact of 1.8E–3*0.34*10*3.78E–10/70*7.2E+09 = 5.71E–04 YLL per year.

The uncertainty is assumed to be a factor of 3, much due to the simple fate model.

7.4.3.2 Disability from cancer

USEPA estimates the lifetime risk for cancer to be 1.8E–3/μg/m^3 of Cd in air. The average duration of cancer morbidity is assumed to be three years.

The fate of a Cd particle is assumed to be the same as an average PM10 particle. The global emission of PM10 is estimated to 63 Tg/yr and the global average PM10 concentration to 61 μg/m^3. Primary emissions contribute by 40% to the total particle mass. This means that 1 kg of Cd emitted to air will result in a concentration of 0.4*61/63E+09 = 3.78E–10 μg/m^3 and an impact of 1.8E–3*3*3.78E–10/70*7.2E+09 = 2.10E–04 personyears per year.

The uncertainty is assumed to be a factor of 3, much due to the simple fate model.

7.4.4 Summary and monetary valuation

A summary of environmental impact factors and monetary impact values are shown in Table 7.8.

It is striking that the monetary impact value is only slightly higher than that for PM2.5. Maybe it has something to do with how risk estimates for Cd and PM2.5 are done. Risk estimates for Cd relies on work environment observations, while the risk for PM2.5 is determined in cohort studies on large population groups. They are risks for different groups under different exposure situations.

Table 7.8. Summary of environmental impact factors and monetary impact values of emission of 1 kg Cd.

Environmental good	Impact indicator	Unit	Pathway	Environmental impact factor	Uncer-tainty	Indicator value ($/unit)	Impact value ($/kg Cd)
human health	YLL	personyears	cancer	2.38E–04	3	107,067	2.55E+01
human health	cancer	personyears	direct exposure	2.10E–04	3	21,413	4.50E+00
						sum	3.00E+01

7.5 Chromium

7.5.1 Specification of the emission

The emission valued is chromium emitted in air anywhere in the world in 2015.

7.5.2 Impact pathways

Chromium (VI) is a potent carcinogen. Part of the chromium found in air is Cr^{6+}. It is unclear if the part of Cr that is Cr^{6+} is stable from the source of Cr, but atmospheric reactions may be influencing the Cr^{6+}/Cr ratio. Cr^{6+} is primarily associated with lung cancer and nasal cancer.

7.5.3 Environmental impact factors

7.5.3.1 YLL from cancer

The cancer risk of Cr^{6+} has been estimated by WHO [22] to be in the order of 4E–02/μg/m^3 expressed as a lifetime unit risk. The hexavalent part of Cr in ambient air is not known in a global context but was estimated by Scott et al. [23] in New Jersey, as an average, to 26%.

The mortality for lung cancer is 83% (National Cancer Institute, http://seer.cancer.gov). The average YLL per fatal lung cancer is estimated to 10 years from studies in Norway and Iran [15, 16]. Iran is assumed to represent the developing world and Norway the developed world.

The fate of a Cr particle is assumed to be the same as an average PM10 particle. The global emission of PM10 is estimated to 63 Tg/yr and the global average PM10 concentration to 61 μg/m^3. Primary particles contribute to 40% of the ambient particle mass. This means that 1 kg of Cr emitted to air will result in a concentration of 0.4*61/63E+09 = 3.78E–10 μg/m^3 and an impact of 0.26*4E–2*0.83*10*3.78E–10/70*7.2E+09 = 3.36E–03 YLL per year.

The uncertainty is assumed to be a factor of 3, much due to the simple fate model.

7.5.3.2 Disability from cancer

The cancer risk of Cr^{6+} has been estimated by WHO [22] to be in the order of 4E–02/μg/m^3 expressed as a lifetime unit risk. The hexavalent part of Cr in ambient air is not known in a global context but was estimated by Scott et al. [23] in New Jersey, as an average, to 26%.

The average disability is estimated to two years.

Table 7.9. Summary of environmental impact factors and monetary impact values of emission of 1 kg Cr.

Environmental good	Impact indicator	Unit	Pathway	Environmental impact factor	Uncertainty	Indicator value ($/unit)	Impact value ($/kg Cr)
human health	YLL	personyears	cancer	3.36E–03	3	107,067	3.59E+02
human health	cancer	personyears	direct exposure	8.09E–04	3	21,413	1.73E+01
						sum	3.77E+02

The fate of a Cr particle is assumed to be the same as an average PM10 particle. The global emission of PM10 is estimated to 63 Tg/yr and the global average PM10 concentration to 61 μg/m^3. Primary particles contribute to 40% of the ambient particle mass. This means that 1 kg of Cr emitted to air will result in a concentration of 0.4*61/63E+09 = 3.78E–10 μg/m^3 and an impact of 0.26*4E–2*2*3.78E–10/70*7.2E+09 = 8.09E–04 personyears per year.

The uncertainty is assumed to be a factor of three, much due to the simple fate model.

7.5.4 Summary and monetary valuation

A summary of environmental impact factors and monetary impact values are shown in Table 7.9.

It is striking that the monetary impact value is only slightly lower than that for PM2.5. Maybe it has something to do with how risk estimates for Cr^{6+} and PM2.5 are done. Risk estimates for Cr^{6+} rely on work environment observations, while the risk for PM2.5 is determined in cohort studies on large population groups. They are risks for different groups under different exposure situations.

References

1. Werner, A.S. et al. *Determination of Condensable and Non-condensable PM2.5 Emissions from Stationary Sources*. 2012.
2. Pei, B., Determination and emission of condensable particulate matter from coal-fired power plants. *Huanjing Kexue/Environmental Science*, 2015. **36**(5): 1544–1549.
3. Cohen, A.J. et al., Estimates and 25-year trends of the global burden of disease attributable to ambient air pollution: an analysis of data from the Global Burden of Diseases Study 2015. *Lancet*, 2017. **389**(May 13): 1907–18.
4. *EDGAR, Emissions Database for Global Athmospheric Research*. 2017, EU JRC.
5. ZHANG, W. et al., Source apportionment for urban PM10 and PM2.5 in the Beijing area. *Chinese Science Bulletin*, 2007. **52**(5): 608–615.
6. Gugamsetty, B. et al., Source Characterization and Apportionment of PM10, PM2.5 and PM0.1 by Using Positive Matrix Factorization. *Aerosol and Air Quality Research*, 2012. **12**: 476–491.
7. Marcazzan, G.M. et al., Source apportionment of PM10 and PM2.5 in Milan (Italy) using receptor modelling. *The Science of the Total Environment*, 2003. **317**: 137–147.
8. Querol, X. et al., *PM10 and PM2.5 source apportionment in the Barcelona Metropolitan area, Catalonia, Spain*. Atmospheric Environment, 2001. **35**: 6407–6419.
9. Masiol, M. et al., Atmospheric Long-term trends (2005–2016) of source apportioned PM2.5 across New York State. *Environment*, 2019. **201**: 110–120.
10. Yang, Y. et al., Short-term and long-term exposures to fine particulate matter constituents and health: A systematic review and meta-analysis. *Environmental Pollution*, 2019. **247**: 874–882.
11. Fuglestvedt, J.S. et al., Transport impacts on atmosphere and climate: Metrics. *Atmospheric Environment*, 2010. **44**(37): 4648–4677.

12. *Some Non-heterocyclic Polycyclic Aromatic Hydrocarbons and Some Related Exposures*, in *IARC Monographs on the Evaluation of Carcinogenic Risks to Humans* 2010, International Agency for Research on Cancer: 150 cours Albert Thomas, 69372 Lyon Cedex 08, France.
13. Shen, H. et al., Global lung cancer risk from PAH exposure highly depends on emission sources and individual susceptibility. *Scientific Reports*, 2014. **4**: 6561.
14. Matsuda, A. and K. Katanoda, Five-Year Relative Survival Rate of Lung Cancer in the USA, Europe and Japan. *Japanese Journal of Clinical Oncology*, 2013. **43**(12): 1287–1288.
15. Brustugun, O.T., B. Møller and Å. Helland, Years of life lost as a measure of cancer burden on a national level. *Br. J. Cancer*, 2014. **111**(5): 1014–1020.
16. Mirzaei, M., M. Mirzadeh and M. Mirzaei, Expected Years of Life Lost Due to Adult Cancer Mortality in Yazd (2004–2010). *Asian Pac. J. Cancer*, Cancer Control in Western Asia Special Issue, 2016. **17**: 101–105.
17. Zhang, Y. and S. Tao, Global atmospheric emission inventory of polycyclic aromatic hydrocarbons (PAHs) for 2004. *Atmospheric Environment*, 2009. **43**(4): 812–819.
18. Who, *Global Burden of Disease 2004 Update: Disability Weights for Diseases and Conditions*. 2004, WHO: Geneva.
19. Lozano, R. et al., Global and regional mortality from 235 causes of death for 20 age groups in 1990 and 2010: a systematic analysis for the Global Burden of Disease Study 2010. *The Lancet*, 2012. **380**(9859): 2095–2128.
20. Vos, T. et al., Years lived with disability (YLDs) for 1160 sequelae of 289 diseases and injuries 1990–2010: a systematic analysis for the Global Burden of Disease Study 2010. *The Lancet*, 2012. **380**(9859): 2163–2196.
21. Friedrich, R., *ESPREME, Estimation of willingness-to-pay to reduce risks of exposure to heavy metals and cost-benefits analysis for reducing heavy metals occurrence in Europe*, in *Activity report*. 2007, EU 6th framework program. p. 58.
22. WHO, *Concise International Chemical Assessment*, in *International Programme on Chemical Safety*. 2013, WHO.
23. Scott, P.K. et al., Background Air Concentrations of Cr(VI) in Hudson County, New Jersey: Implications for setting Health based Standards for Cr(VI) in Soil. *J. Air & Waste Manage. Assoc.*, 1997. **47**: 592–600

8 Emissions of Halogenated Organic Compounds to Air

The emissions for which environmental impacts are valued here are emissions anywhere in the world, emitted as gases and, where there are no local acute effects.

Halogenated organic compounds may cause depletion of the stratospheric ozone layer and contribute to climate change.

In the modelling of monetary environmental impact values, CO_2 is used as reference for climate impacts and CFC-11 is used as reference for impacts via ozone depletion.

Having made valuation of CO_2 (see Sec. 6.1) and CFC-11, values for individual halogenated organic compounds may be obtained through multiplication by its GWP100 and its ozone depletion potential (ODP).

8.1 CFC-11

8.1.1 Specification of the emission

CFC-11 is a short name for trichlorofluoromethane. It was earlier used frequently for refrigeration and foaming but was banned after the International Montreal Agreement due to its negative effect on the stratospheric ozone layer.

The emission valued is emission of CFC-11 anywhere in the world at any source strength during 2015.

8.1.2 Impact pathways

CFC-11 and many other halogenated organic compounds have long residence times in the air and for a long time an extra kg of CFC-11 will contribute to decreased stratospheric ozone layer. The environmental impact of an emission will depend on the future emission scenario. With low emissions the impact of an extra kg emitted will be small, with high emissions it may lead to severe damages.

CFC-11 also is a climate gas. Its GWP100 is 5332.

8.1.3 Environmental impact factors

8.1.3.1 YLL from cancer via depletion of stratospheric ozone

An emission scenario based on the Copenhagen amendment to the Montreal protocol is used. For that scenario, WHO [1] estimates an average incidence of around 50 cases per million inhabitants per year until 2100, when excess cancer incidence cease. There are three types of skin cancer: melanoma, squamous cell carcinoma (SCC), and basal cell carcinoma (BCC). Mortality in skin cancer is low, about 12% for melanoma and 0.2% for SCC and BCC. If applying the ratio of predicted cancer incidences for melanoma vs. SCC + BCC in the USA (2.7% of all cases) [2] to global conditions, the global mortality rate until 2100 will be 50/1E+06*7.2E+09*88* (0.12*0.027+0.002*0.973) = 1.64E+05 cases of excess mortality.

The average shortening of life in case of fatal melanoma is 14.1 years in England [3] and 17.1 years in Norway [4]. An best estimate of 15 years is assumed. This means that there are 2.46E+06 YLL caused by ozone depletion.

The total emission of ozone depletion substances to 2100 is estimated from a scenario made by Ravishankara et al. [5] to 22000 kton or 2.2E+10 kg CFC-11 equivalents. An emission of 1 kg CFC-11 equivalents therefore contributes with 2.30E+06/2.2E+10 YLL = 1.05E-04 YLL.

The long-time perspective introduces a large uncertainty in the estimation of the environmental impact factor. A factor of 3 is assumed.

8.1.3.2 Skin cancer via depletion of stratospheric ozone

The emission scenario based on the Copenhagen amendment to the Montreal protocol is used. For that, WHO estimates an average incidence of around 50 cases per million inhabitants per year until 2100, when excess cancer incidence ceases [1]. There are three types of skin cancer:

melanoma, squamous cell carcinoma (SCC), and basal cell carcinoma (BCC). The duration of disability due to skin cancers (diagnosis and primary therapy) is estimated to be rather short, 1 month. We thus get 50/1000000*7.2E+09/12 = 3.00E+04 personyears of cancer disability.

The total emission of ozone depletion substances to 2100 is estimated from a scenario made by Ravishankara et al. [5] to 22000 kton or 2.2E+10 kg CFC-11 equivalents. An emission of 1 kg CFC-11 equivalents therefore contributes with 2.92E+04/2.2E+10 YLL = 1.33E–06 YLL/kg CFC-11.

The uncertain estimation of future emissions as well as the duration of the disability motivate an uncertainty factor of 3.

8.1.3.3 Low vision via depletion of stratospheric ozone

USEPA estimated 2010 that in the US there will be 457 500 cataract cases until 2100 with the 1997 amendments to the Montreal protocol. Considering that there is a bias towards the northern part of the US in cataract frequency, the US figures cannot directly be scaled up to represent global conditions, but lacking better assessments, it is here assumed to be relevant for people living in similar latitudes, i.e., about 2 billion people. This means that the global cataract incidence to 2100 would be 457500/313*2000 = 2.92E+06 cases. Most cataract cases are treated by operation, but time of operation could be a year or so. An average disabled time of 2 years is assumed, in order to include those cases that will not be operated. The total extent of the impact is thus 5.85E+06 personyears.

The total emission of ozone depletion substances until 2100 is estimated from a scenario made by Ravishankara et al. [5] to 22000 kton or 2.2E+10 kg CFC-11 equivalents. An emission of 1 kg CFC-11 equivalents therefore contributes with 5.85E+06/2.2E+10 personyears = 2.66E–04personyears of low vision per kg of CFC-11.

As for skin cancer, the uncertainty is large, a factor of 3 is assumed.

8.2 Summary and Monetary Valuation of Emissions of Halogenated Organic Compounds

GWP-100, environmental impact factors for halogenated compounds with respect to ozone depletion (ODP), and total monetary values for ozone depletion and climate change impacts are shown in Table 8.1.

References to GWP and ODP values are given in the table. GWP values (a) are from IPCC AR5 WGI Table 8 SM16, and (d) from IPCC 1994 [6]. (b), (c) are average values for CFCs and HCFCs. For (e) the value is

Table 8.1. Environmental impact factor and monetary impact values for several halogenated organic compounds.

Substance	GWP–100	Ref.	ODP	Ref.	YLL, ozone depletion	YLL, climate change	Skin cancer, ozone depletion	Low vision, ozone depletion	Under-nutrition, climate change	Working capacity, climate change	Diarrhea, climate change	Crop, climate change	Meat, climate change	Wood, climate change	Drinking water, climate change	Bio-diversity, climate change	Impact value ($)
CFC:s																	
CFC-11	5332	a	1	f	1.05E–04	1.00E–02	1.33E–06	2.66E–04	9.28E–03	2.41E+01	1.43E–06	5.40E+01	1.98E+00	0	4.83E+00	9.01E–13	1.90E+03
CFC-12	11547	a	1	f	1.05E–04	2.17E–02	1.33E–06	2.66E–04	2.01E–02	5.23E+01	3.10E–06	1.17E+02	4.29E+00	0	1.05E+01	1.95E–12	4.10E+03
CFC-13	15451	a	1	f	1.05E–04	2.90E–02	1.33E–06	2.66E–04	2.69E–02	7.00E+01	4.15E–06	1.56E+02	5.74E+00	0	1.40E+01	2.61E–12	5.48E+03
CFC-111	9500	b	1	f	1.05E–04	1.79E–02	1.33E–06	2.66E–04	1.65E–02	4.30E+01	2.55E–06	9.62E+01	3.53E+00	0	8.61E+00	1.61E–12	3.38E+03
CFC-112	9500	b	1	f	1.05E–04	1.79E–02	1.33E–06	2.66E–04	1.65E–02	4.30E+01	2.55E–06	9.62E+01	3.53E+00	0	8.61E+00	1.61E–12	3.38E+03
CFC-113	6586	a	1	f	1.05E–04	1.24E–02	1.33E–06	2.66E–04	1.15E–02	2.98E+01	1.77E–06	6.67E+01	2.45E+00	0	5.97E+00	1.11E–12	2.35E+03
CFC-114	9615	a	1	f	1.05E–04	1.81E–02	1.33E–06	2.66E–04	1.67E–02	4.35E+01	2.58E–06	9.73E+01	3.57E+00	0	8.71E+00	1.62E–12	3.42E+03
CFC-115	8516	a	0.44	f	4.62E–05	1.60E–02	5.85E–07	1.17E–04	1.48E–02	3.86E+01	2.29E–06	8.62E+01	3.17E+00	0	7.71E+00	1.44E–12	3.02E+03
CFC-211	9500	b	1	g	1.05E–04	1.79E–02	1.33E–06	2.66E–04	1.65E–02	4.30E+01	2.55E–06	9.62E+01	3.53E+00	0	8.61E+00	1.61E–12	3.38E+03
CFC-212	9500	b	1	g	1.05E–04	1.79E–02	1.33E–06	2.66E–04	1.65E–02	4.30E+01	2.55E–06	9.62E+01	3.53E+00	0	8.61E+00	1.61E–12	3.38E+03
CFC-213	9500	b	1	g	1.05E–04	1.79E–02	1.33E–06	2.66E–04	1.65E–02	4.30E+01	2.55E–06	9.62E+01	3.53E+00	0	8.61E+00	1.61E–12	3.38E+03
CFC-214	9500	b	1	g	1.05E–04	1.79E–02	1.33E–06	2.66E–04	1.65E–02	4.30E+01	2.55E–06	9.62E+01	3.53E+00	0	8.61E+00	1.61E–12	3.38E+03
CFC-215	9500	b	1	g	1.05E–04	1.79E–02	1.33E–06	2.66E–04	1.65E–02	4.30E+01	2.55E–06	9.62E+01	3.53E+00	0	8.61E+00	1.61E–12	3.38E+03
CFC-216	9500	b	1	g	1.05E–04	1.79E–02	1.33E–06	2.66E–04	1.65E–02	4.30E+01	2.55E–06	9.62E+01	3.53E+00	0	8.61E+00	1.61E–12	3.38E+03
CFC-217	9500	b	1	g	1.05E–04	1.79E–02	1.33E–06	2.66E–04	1.65E–02	4.30E+01	2.55E–06	9.62E+01	3.53E+00	0	8.61E+00	1.61E–12	3.38E+03
HCFC:s																	
HCFC-21	179	a	0.04	h	4.20E–06	3.37E–04	5.32E–08	1.06E–05	3.11E–04	8.11E–01	4.81E–08	1.81E+00	6.66E–02	0	1.62E–01	3.03E–14	6.40E+01
HCFC-22	2106	a	0.055	h	5.78E–06	3.96E–03	7.32E–08	1.46E–05	3.66E–03	9.54E+00	5.66E–07	2.13E+01	7.83E–01	0	1.91E+00	3.56E–13	7.46E+02

HCFC-31	700	c	0.02		2.10E–06	1.32E–03	2.66E–08	5.32E–06	1.22E–03	3.17E+00	1.88E–07	7.09E+00	2.60E–01	0	6.34E–01	1.18E–13	2.48E+02
HCFC-121	700	c	0.025	h	2.63E–06	1.32E–03	3.33E–08	6.65E–06	1.22E–03	3.17E+00	1.88E–07	7.09E+00	2.60E–01	0	6.34E–01	1.18E–13	2.48E+02
HCFC-122	72	a	0.05	h	5.25E–06	1.35E–04	6.65E–08	1.33E–05	1.25E–04	3.26E–01	1.93E–08	7.29E–01	2.68E–02	0	6.52E–02	1.22E–14	2.63E+01
HCFC-122a	312	a	0.05	h	5.25E–06	5.87E–04	6.65E–08	1.33E–05	5.43E–04	1.41E+00	8.38E–08	3.16E+00	1.16E–01	0	2.83E–01	5.27E–14	1.11E+02
HCFC-123	96	a	0.04	h	4.20E–06	1.80E–04	5.32E–08	1.06E–05	1.67E–04	4.35E–01	2.58E–08	9.72E–01	3.57E–02	0	8.70E–02	1.62E–14	3.46E+01
HCFC-123a	447	a	0.04	h	4.20E–06	8.40E–04	5.32E–08	1.06E–05	7.78E–04	2.02E+00	1.20E–07	4.53E+00	1.66E–01	0	4.05E–01	7.55E–14	1.59E+02
HCFC-124	635	a	0.03	h	3.15E–06	1.19E–03	3.99E–08	7.98E–06	1.10E–03	2.88E+00	1.71E–07	6.43E+00	2.36E–01	0	5.75E–01	1.07E–13	2.25E+02
HCFC-131	700	c	0.028	h	2.94E–06	1.32E–03	3.72E–08	7.45E–06	1.22E–03	3.17E+00	1.88E–07	7.09E+00	2.60E–01	0	6.34E–01	1.18E–13	2.48E+02
HCFC-132c	409	a	0.029	h	3.05E–06	7.69E–04	3.86E–08	7.71E–06	7.12E–04	1.85E+00	1.10E–07	4.14E+00	1.52E–01	0	3.70E–01	6.91E–14	1.45E+02
HCFC-133	700	c	0.04	h	4.20E–06	1.32E–03	5.32E–08	1.06E–05	1.22E–03	3.17E+00	1.88E–07	7.09E+00	2.60E–01	0	6.34E–01	1.18E–13	2.48E+02
HCFC-141	700	c	0.037	h	3.89E–06	1.32E–03	4.92E–08	9.84E–06	1.22E–03	3.17E+00	1.88E–07	7.09E+00	2.60E–01	0	6.34E–01	1.18E–13	2.48E+02
HCFC-141b	938	a	0.11	h	1.16E–05	1.76E–03	1.46E–07	2.93E–05	1.63E–03	4.25E+00	2.52E–07	9.50E+00	3.49E–01	0	8.50E–01	1.59E–13	3.34E+02
HCFC-142	700	c	0.039	h	4.10E–06	1.32E–03	5.19E–08	1.04E–05	1.22E–03	3.17E+00	1.88E–07	7.09E+00	2.60E–01	0	6.34E–01	1.18E–13	2.48E+02
HCFC-142b	2345	a	0.065	h	6.83E–06	4.41E–03	8.65E–08	1.73E–05	4.08E–03	1.06E+01	6.30E–07	2.37E+01	8.72E–01	0	2.12E+00	3.96E–13	8.31E+02
HCFC-151	700	c	0.04	h	4.20E–06	1.32E–03	5.32E–08	1.06E–05	1.22E–03	3.17E+00	1.88E–07	7.09E+00	2.60E–01	0	6.34E–01	1.18E–13	2.48E+02
HCFC-221	700	c	0.042	h	4.41E–06	1.32E–03	5.59E–08	1.12E–05	1.22E–03	3.17E+00	1.88E–07	7.09E+00	2.60E–01	0	6.34E–01	1.18E–13	2.48E+02
HCFC-222	700	c	0.05	h	5.25E–06	1.32E–03	6.65E–08	1.33E–05	1.22E–03	3.17E+00	1.88E–07	7.09E+00	2.60E–01	0	6.34E–01	1.18E–13	2.49E+02
HCFC-223	700	c	0.04	h	4.20E–06	1.32E–03	5.32E–08	1.06E–05	1.22E–03	3.17E+00	1.88E–07	7.09E+00	2.60E–01	0	6.34E–01	1.18E–13	2.48E+02
HCFC-224	700	c	0.05	h	5.25E–06	1.32E–03	6.65E–08	1.33E–05	1.22E–03	3.17E+00	1.88E–07	7.09E+00	2.60E–01	0	6.34E–01	1.18E–13	2.49E+02
HCFC-225	700	c	0.04	h	4.20E–06	1.32E–03	5.32E–08	1.06E–05	1.22E–03	3.17E+00	1.88E–07	7.09E+00	2.60E–01	0	6.34E–01	1.18E–13	2.48E+02
HCFC-225ca	155	a	0.025	h	2.63E–06	2.91E–04	3.33E–08	6.65E–06	2.70E–04	7.02E–01	4.17E–08	1.57E+00	5.76E–02	0	1.40E–01	2.62E–14	5.53E+01
HCFC-225cb	633	a	0.033	h	3.47E–06	1.19E–03	4.39E–08	8.78E–06	1.10E–03	2.87E+00	1.70E–07	6.41E+00	2.35E–01	0	5.73E–01	1.07E–13	2.25E+02

Table 8.1 contd. …

...Table 8.1 contd.

Substance	GWP-100	Ref.	ODP	Ref.	YLL, ozone depletion	YLL, climate change	Skin cancer, ozone depletion	Low vision, ozone depletion	Under-nutrition, climate change	Working capacity, climate change	Diarrhea, climate change	Crop, climate change	Meat, climate change	Wood, climate change	Drinking water, climate change	Bio-diversity, climate change	Impact value ($)
HCFC-226	700	c	0.06	h	6.30E–06	1.32E–03	7.98E–08	1.60E–05	1.22E–03	3.17E+00	1.88E–07	7.09E+00	2.60E–01	0	6.34E–01	1.18E–13	2.49E+02
HCFC-231	700	c	0.07	h	7.35E–06	1.32E–03	9.31E–08	1.86E–05	1.22E–03	3.17E+00	1.88E–07	7.09E+00	2.60E–01	0	6.34E–01	1.18E–13	2.49E+02
HCFC-232	700	c	0.054	h	5.67E–06	1.32E–03	7.18E–08	1.44E–05	1.22E–03	3.17E+00	1.88E–07	7.09E+00	2.60E–01	0	6.34E–01	1.18E–13	2.49E+02
HCFC-233	700	c	0.118	h	1.24E–05	1.32E–03	1.57E–07	3.14E–05	1.22E–03	3.17E+00	1.88E–07	7.09E+00	2.60E–01	0	6.34E–01	1.18E–13	2.50E+02
HCFC-234	700	c	0.14	h	1.47E–05	1.32E–03	1.86E–07	3.72E–05	1.22E–03	3.17E+00	1.88E–07	7.09E+00	2.60E–01	0	6.34E–01	1.18E–13	2.50E+02
HCFC-235	700	c	0.27	h	2.84E–05	1.32E–03	3.59E–07	7.18E–05	1.22E–03	3.17E+00	1.88E–07	7.09E+00	2.60E–01	0	6.34E–01	1.18E–13	2.52E+02
HCFC-241	700	c	0.047	h	4.94E–06	1.32E–03	6.25E–08	1.25E–05	1.22E–03	3.17E+00	1.88E–07	7.09E+00	2.60E–01	0	6.34E–01	1.18E–13	2.49E+02
HCFC-242	700	c	0.067	h	7.04E–06	1.32E–03	8.91E–08	1.78E–05	1.22E–03	3.17E+00	1.88E–07	7.09E+00	2.60E–01	0	6.34E–01	1.18E–13	2.49E+02
HCFC-243	700	c	0.63	h	6.62E–05	1.32E–03	8.38E–07	1.68E–04	1.22E–03	3.17E+00	1.88E–07	7.09E+00	2.60E–01	0	6.34E–01	1.18E–13	2.58E+02
HCFC-244	700	c	0.074	h	7.77E–06	1.32E–03	9.84E–08	1.97E–05	1.22E–03	3.17E+00	1.88E–07	7.09E+00	2.60E–01	0	6.34E–01	1.18E–13	2.49E+02
HCFC-251	700	c	0.005	h	5.25E–07	1.32E–03	6.65E–09	1.33E–06	1.22E–03	3.17E+00	1.88E–07	7.09E+00	2.60E–01	0	6.34E–01	1.18E–13	2.48E+02
HCFC-252	700	c	0.022	h	2.31E–06	1.32E–03	2.93E–08	5.85E–06	1.22E–03	3.17E+00	1.88E–07	7.09E+00	2.60E–01	0	6.34E–01	1.18E–13	2.48E+02
HCFC-253	700	c	0.016	h	1.68E–06	1.32E–03	2.13E–08	4.26E–06	1.22E–03	3.17E+00	1.88E–07	7.09E+00	2.60E–01	0	6.34E–01	1.18E–13	2.48E+02
HCFC-261	700	c	0.011	h	1.16E–06	1.32E–03	1.46E–08	2.93E–06	1.22E–03	3.17E+00	1.88E–07	7.09E+00	2.60E–01	0	6.34E–01	1.18E–13	2.48E+02
HCFC-262	700	c	0.011	h	1.16E–06	1.32E–03	1.46E–08	2.93E–06	1.22E–03	3.17E+00	1.88E–07	7.09E+00	2.60E–01	0	6.34E–01	1.18E–13	2.48E+02
HCFC-23	93	d	0.02	h	2.10E–06	1.75E–04	2.66E–08	5.32E–06	1.62E–04	4.21E–01	2.50E–08	9.41E–01	3.46E–02	0	8.42E–02	1.57E–14	3.32E+01
HCFC-24	480	d	0.022	h	2.31E–06	9.02E–04	2.93E–08	5.85E–06	8.35E–04	2.17E+00	1.29E–07	4.86E+00	1.78E–01	0	4.35E–01	8.11E–14	1.70E+02
HCFC-25	630	d	0.11	h	1.16E–05	1.18E–03	1.46E–07	2.93E–05	1.10E–03	2.85E+00	1.69E–07	6.38E+00	2.34E–01	0	5.71E–01	1.06E–13	2.25E+02
HCFC-26	2000	d	0.065	h	6.83E–06	3.76E–03	8.65E–08	1.73E–05	3.48E–03	9.06E+00	5.37E–07	2.02E+01	7.44E–01	0	1.81E+00	3.38E–13	7.09E+02

HCFC-27	170	d	0.025	h	2.63E–06	3.20E–04	3.33E–08	6.65E–06	2.96E–04	7.70E–01	4.57E–08	1.72E+00	6.32E–02	0	1.54E–01	2.87E–14	6.06E+01
HCFC-28	530	d	0.033	h	3.47E–06	9.96E–04	4.39E–08	8.78E–06	9.22E–04	2.40E+00	1.42E–07	5.37E+00	1.97E–01	0	4.80E–01	8.96E–14	1.88E+02
HCFC-271	700	c	0.015	h	1.58E–06	1.32E–03	2.00E–08	3.99E–06	1.22E–03	3.17E+00	1.88E–07	7.09E+00	2.60E–01	0	6.34E–01	1.18E–13	2.48E+02
(E)-1-Chloro-3,3,3-trifluoroprop-1-ene	2	a			0.00E+00	3.76E–06	0.00E+00	0.00E+00	3.48E–06	9.06E–03	5.37E-10	2.02E–02	7.44E–04	0	1.81E–03	3.38E–16	7.08E–01
HFC-23	13856	a			0.00E+00	2.60E–02	0.00E+00	0.00E+00	2.41E–02	6.28E+01	3.72E–06	1.40E+02	5.15E+00	0	1.26E+01	2.34E–12	4.90E+03
HFC-32	817	a			0.00E+00	1.54E–03	0.00E+00	0.00E+00	1.42E–03	3.70E+00	2.20E–07	8.27E+00	3.04E–01	0	7.40E–01	1.38E–13	2.89E+02
HFC-41	141	a			0.00E+00	2.65E–04	0.00E+00	0.00E+00	2.45E–04	6.39E–01	3.79E–08	1.43E+00	5.24E–02	0	1.28E–01	2.38E–14	4.99E+01
HFC-43-10mee	1600	d			0.00E+00	3.01E–03	0.00E+00	0.00E+00	2.78E–03	7.25E+00	4.30E–07	1.62E+01	5.95E–01	0	1.45E+00	2.70E–13	5.66E+02
HFC-125	3691	a			0.00E+00	6.94E–03	0.00E+00	0.00E+00	6.42E–03	1.67E+01	9.92E–07	3.74E+01	1.37E+00	0	3.34E+00	6.24E–13	1.31E+03
HFC-134	1337	a			0.00E+00	2.51E–03	0.00E+00	0.00E+00	2.33E–03	6.06E+00	3.59E–07	1.35E+01	4.97E–01	0	1.21E+00	2.26E–13	4.73E+02
HFC-134a	1549	a			0.00E+00	2.91E–03	0.00E+00	0.00E+00	2.70E–03	7.02E+00	4.16E–07	1.57E+01	5.76E–01	0	1.40E+00	2.62E–13	5.48E+02
HFC-143	397	a			0.00E+00	7.46E–04	0.00E+00	0.00E+00	6.91E–04	1.80E+00	1.07E–07	4.02E+00	1.48E–01	0	3.60E–01	6.71E–14	1.41E+02
HFC-143a	5508	a			0.00E+00	1.04E–02	0.00E+00	0.00E+00	9.58E–03	2.49E+01	1.48E–06	5.58E+01	2.05E+00	0	4.99E+00	9.31E–13	1.95E+03
HFC-152	20	a			0.00E+00	3.76E–05	0.00E+00	0.00E+00	3.48E–05	9.06E–02	5.37E–09	2.02E–01	7.44E–03	0	1.81E–02	3.38E–15	7.08E+00
HFC-152a	167	a			0.00E+00	3.14E–04	0.00E+00	0.00E+00	2.91E–04	7.56E–01	4.49E–08	1.69E+00	6.21E–02	0	1.51E–01	2.82E–14	5.91E+01
HFC-161	4	a			0.00E+00	7.52E–06	0.00E+00	0.00E+00	6.96E–06	1.81E–02	1.07E–09	4.05E–02	1.49E–03	0	3.62E–03	6.76E–16	1.42E+00
HFC-227ca	3077	a			0.00E+00	5.78E–03	0.00E+00	0.00E+00	5.35E–03	1.39E+01	8.27E–07	3.11E+01	1.14E+00	0	2.79E+00	5.20E–13	1.09E+03
HFC-227ea	3860	a			0.00E+00	7.26E–03	0.00E+00	0.00E+00	6.72E–03	1.75E+01	1.04E–06	3.91E+01	1.44E+00	0	3.50E+00	6.52E–13	1.37E+03
HFC-236cb	1438	a			0.00E+00	2.70E–03	0.00E+00	0.00E+00	2.50E–03	6.51E+00	3.86E–07	1.46E+01	5.35E–01	0	1.30E+00	2.43E–13	5.09E+02
HFC-236ea	1596	a			0.00E+00	3.00E–03	0.00E+00	0.00E+00	2.78E–03	7.23E+00	4.29E–07	1.62E+01	5.93E–01	0	1.45E+00	2.70E–13	5.65E+02
HFC-236fa	8998	a			0.00E+00	1.69E–02	0.00E+00	0.00E+00	1.57E–02	4.08E+01	2.42E–06	9.11E+01	3.35E+00	0	8.15E+00	1.52E–12	3.18E+03

Table 8.1 contd. …

...Table 8.1 contd.

Substance	GWP–100	Ref.	ODP	Ref.	YLL, ozone depletion	YLL, climate change	Skin cancer, ozone depletion	Low vision, ozone depletion	Under-nutrition, climate change	Working capacity, climate change	Diarrhea, climate change	Crop, climate change	Meat, climate change	Wood, climate change	Drinking water, climate change	Bio-diversity, climate change	Impact value ($)
HFC-245ca	863	a			0.00E+00	1.62E–03	0.00E+00	0.00E+00	1.50E–03	3.91E+00	2.32E–07	8.74E+00	3.21E–01	0	7.82E–01	1.46E–13	3.05E+02
HFC-245cb	5298	a			0.00E+00	9.96E–03	0.00E+00	0.00E+00	9.22E–03	2.40E+01	1.42E–06	5.36E+01	1.97E+00	0	4.80E+00	8.95E–13	1.88E+03
HFC-245ea	285	a			0.00E+00	5.36E–04	0.00E+00	0.00E+00	4.96E–04	1.29E+00	7.66E–08	2.89E+00	1.06E–01	0	2.58E–01	4.82E–14	1.01E+02
HFC-245eb	352	a			0.00E+00	6.62E–04	0.00E+00	0.00E+00	6.12E–04	1.59E+00	9.46E–08	3.56E+00	1.31E–01	0	3.19E–01	5.95E–14	1.25E+02
HFC-245fa	1032	a			0.00E+00	1.94E–03	0.00E+00	0.00E+00	1.80E–03	4.67E+00	2.77E–07	1.04E+01	3.84E–01	0	9.35E–01	1.74E–13	3.65E+02
HFC-263fb	92	a			0.00E+00	1.73E–04	0.00E+00	0.00E+00	1.60E–04	4.17E–01	2.47E–08	9.31E–01	3.42E–02	0	8.33E–02	1.55E–14	3.26E+01
HFC-272ca	175	a			0.00E+00	3.29E–04	0.00E+00	0.00E+00	3.05E–04	7.93E–01	4.70E–08	1.77E+00	6.51E–02	0	1.59E–01	2.96E–14	6.19E+01
HFC-329p	2742	a			0.00E+00	5.15E–03	0.00E+00	0.00E+00	4.77E–03	1.24E+01	7.37E–07	2.78E+01	1.02E+00	0	2.48E+00	4.63E–13	9.70E+02
HFC-365mfc	966	a			0.00E+00	1.82E–03	0.00E+00	0.00E+00	1.68E–03	4.38E+00	2.60E–07	9.78E+00	3.59E–01	0	8.75E–01	1.63E–13	3.42E+02
HFC-43-10mee	1952	a			0.00E+00	3.67E–03	0.00E+00	0.00E+00	3.40E–03	8.84E+00	5.25E–07	1.98E+01	7.26E–01	0	1.77E+00	3.30E–13	6.91E+02
HFC-1132a	1	a			0.00E+00	1.88E–06	0.00E+00	0.00E+00	1.74E–06	4.53E–03	2.69E-10	1.01E–02	3.72E–04	0	9.06E–04	1.69E–16	3.54E–01
HFC-1141	1	a			0.00E+00	1.88E–06	0.00E+00	0.00E+00	1.74E–06	4.53E–03	2.69E-10	1.01E–02	3.72E–04	0	9.06E–04	1.69E–16	3.54E–01
(Z)-HFC-1225ye	1	a			0.00E+00	1.88E–06	0.00E+00	0.00E+00	1.74E–06	4.53E–03	2.69E-10	1.01E–02	3.72E–04	0	9.06E–04	1.69E–16	3.54E–01
(E)-HFC-1225ye	1	a			0.00E+00	1.88E–06	0.00E+00	0.00E+00	1.74E–06	4.53E–03	2.69E-10	1.01E–02	3.72E–04	0	9.06E–04	1.69E–16	3.54E–01
(Z)-HFC-1234ze	1	a			0.00E+00	1.88E–06	0.00E+00	0.00E+00	1.74E–06	4.53E–03	2.69E-10	1.01E–02	3.72E–04	0	9.06E–04	1.69E–16	3.54E–01
HFC-1234yf	1	a			0.00E+00	1.88E–06	0.00E+00	0.00E+00	1.74E–06	4.53E–03	2.69E-10	1.01E–02	3.72E–04	0	9.06E–04	1.69E–16	3.54E–01
(E)-HFC-1234ze	1	a			0.00E+00	1.88E–06	0.00E+00	0.00E+00	1.74E–06	4.53E–03	2.69E-10	1.01E–02	3.72E–04	0	9.06E–04	1.69E–16	3.54E–01
(Z)-HFC-1336	2	a			0.00E+00	3.76E–06	0.00E+00	0.00E+00	3.48E–06	9.06E–03	5.37E-10	2.02E–02	7.44E–04	0	1.81E–03	3.38E–16	7.08E–01
HFC-1243zf	1	a			0.00E+00	1.88E–06	0.00E+00	0.00E+00	1.74E–06	4.53E–03	2.69E-10	1.01E–02	3.72E–04	0	9.06E–04	1.69E–16	3.54E–01

HFC-1345zfc	1	a			0.00E+00	1.88E–06	0.00E+00	0.00E+00	1.74E–06	4.53E–03	2.69E-10	1.01E–02	3.72E–04	0	9.06E–04	1.69E–16	3.54E–01
3,3,4,4,5,5,6,6,6-Nonafluorohex-1-ene	1	a			0.00E+00	1.88E–06	0.00E+00	0.00E+00	1.74E–06	4.53E–03	2.69E-10	1.01E–02	3.72E–04	0	9.06E–04	1.69E–16	3.54E–01
3,3,4,4,5,5,6,6,7,7,8,8,8-Tridecafluorooct-1-ene	1	a			0.00E+00	1.88E–06	0.00E+00	0.00E+00	1.74E–06	4.53E–03	2.69E-10	1.01E–02	3.72E–04	0	9.06E–04	1.69E–16	3.54E–01
3,3,4,4,5,5,6,6,7,7,8,8,9,9,10,10,10-Heptadecafluorodec-1-ene	1	a			0.00E+00	1.88E–06	0.00E+00	0.00E+00	1.74E–06	4.53E–03	2.69E-10	1.01E–02	3.72E–04	0	9.06E–04	1.69E–16	3.54E–01
CCl4	2019	a	0.73	f	7.67E–05	3.80E–03	9.71E–07	1.94E–04	3.51E–03	9.14E+00	5.43E–07	2.04E+01	7.51E–01	0	1.83E+00	3.41E–13	7.26E+02
Methyl Chloroform (C2H3Cl3) 1,1,1-trichloroethane	160	a	0.12	f	1.26E–05	3.01E–04	1.60E–07	3.19E–05	2.78E–04	7.25E–01	4.30E–08	1.62E+00	5.95E–02	0	1.45E–01	2.70E–14	5.86E+01
Bromocarbons																	
Methyl bromide	3	a	0.51	f	5.36E–05	5.64E–06	6.78E–07	1.36E–04	5.22E–06	1.36E–02	8.06E-10	3.04E–02	1.12E–03	0	2.72E–03	5.07E–16	9.27E+00
Methylene bromide	1	a			0.00E+00	1.88E–06	0.00E+00	0.00E+00	1.74E–06	4.53E–03	2.69E-10	1.01E–02	3.72E–04	0	9.06E–04	1.69E–16	3.54E–01
Halon-1201	454	a	0.74		7.77E–05	8.54E–04	9.84E–07	1.97E–04	7.90E–04	2.06E+00	1.22E–07	4.60E+00	1.69E–01	0	4.11E–01	7.67E–14	1.73E+02
Halon-1202	280	a	0.62	i	6.51E–05	5.26E–04	8.25E–07	1.65E–04	4.87E–04	1.27E+00	7.52E–08	2.83E+00	1.04E–01	0	2.54E–01	4.73E–14	1.09E+02
Halon-1211	2070	a	7.1	f	7.46E–04	3.89E–03	9.44E–06	1.89E–03	3.60E–03	9.38E+00	5.56E–07	2.10E+01	7.70E–01	0	1.88E+00	3.50E–13	8.47E+02
Halon-1301	7154	a	16	f	1.68E–03	1.34E–02	2.13E–05	4.26E–03	1.24E–02	3.24E+01	1.92E–06	7.24E+01	2.66E+00	0	6.48E+00	1.21E–12	2.79E+03
Halon-2301	210	a			0.00E+00	3.95E–04	0.00E+00	0.00E+00	3.65E–04	9.51E–01	5.64E–08	2.13E+00	7.81E–02	0	1.90E–01	3.55E–14	7.43E+01
Halon-2311/Halothane	50	a			0.00E+00	9.40E–05	0.00E+00	0.00E+00	8.70E–05	2.26E–01	1.34E–08	5.06E–01	1.86E–02	0	4.53E–02	8.45E–15	1.77E+01
Halon-2401	223	a			0.00E+00	4.19E–04	0.00E+00	0.00E+00	3.88E–04	1.01E+00	5.99E–08	2.26E+00	8.29E–02	0	2.02E–01	3.77E–14	7.89E+01
Halon-2402	1734	a	11.5	f	1.21E–03	3.26E–03	1.53E–05	3.06E–03	3.02E–03	7.85E+00	4.66E–07	1.76E+01	6.45E–01	0	1.57E+00	2.93E–13	7.99E+02

Table 8.1 contd. …

...Table 8.1 contd.

Substance	GWP–100	Ref.	ODP	Ref.	YLL, ozone depletion	YLL, climate change	Skin cancer, ozone depletion	Low vision, ozone depletion	Under-nutrition, climate change	Working capacity, climate change	Diarrhea, climate change	Crop, climate change	Meat, climate change	Wood, climate change	Drinking water, climate change	Bio-diversity, climate change	Impact value ($)
Fluorocarbons																	
Nitrogen trifluoride	17885	a			0.00E+00	3.36E–02	0.00E+00	0.00E+00	3.11E–02	8.10E+01	4.81E–06	1.81E+02	6.65E+00	0	1.62E+01	3.02E–12	6.33E+03
Sulphur hexafluoride	26087	a			0.00E+00	4.90E–02	0.00E+00	0.00E+00	4.54E–02	1.18E+02	7.01E–06	2.64E+02	9.70E+00	0	2.36E+01	4.41E–12	9.23E+03
(Trifluoromethyl)sulphur pentafluoride	19396	a			0.00E+00	3.65E–02	0.00E+00	0.00E+00	3.37E–02	8.78E+01	5.21E–06	1.96E+02	7.21E+00	0	1.76E+01	3.28E–12	6.86E+03
Sulphuryl fluoride	4732	a			0.00E+00	8.90E–03	0.00E+00	0.00E+00	8.23E–03	2.14E+01	1.27E–06	4.79E+01	1.76E+00	0	4.29E+00	8.00E–13	1.67E+03
PFC-14	7349	a			0.00E+00	1.38E–02	0.00E+00	0.00E+00	1.28E–02	3.33E+01	1.97E–06	7.44E+01	2.73E+00	0	6.66E+00	1.24E–12	2.60E+03
PFC-116	12340	a			0.00E+00	2.32E–02	0.00E+00	0.00E+00	2.15E–02	5.59E+01	3.32E–06	1.25E+02	4.59E+00	0	1.12E+01	2.09E–12	4.37E+03
PFC-c216	10208	a			0.00E+00	1.92E–02	0.00E+00	0.00E+00	1.78E–02	4.62E+01	2.74E–06	1.03E+02	3.80E+00	0	9.25E+00	1.73E–12	3.61E+03
PFC-218	9878	a			0.00E+00	1.86E–02	0.00E+00	0.00E+00	1.72E–02	4.47E+01	2.65E–06	1.00E+02	3.67E+00	0	8.95E+00	1.67E–12	3.50E+03
PFC-318	10592	a			0.00E+00	1.99E–02	0.00E+00	0.00E+00	1.84E–02	4.80E+01	2.85E–06	1.07E+02	3.94E+00	0	9.59E+00	1.79E–12	3.75E+03
PFC-31-10	10213	a			0.00E+00	1.92E–02	0.00E+00	0.00E+00	1.78E–02	4.63E+01	2.74E–06	1.03E+02	3.80E+00	0	9.25E+00	1.73E–12	3.61E+03
Perfluor-ocyclopentene	2	a			0.00E+00	3.76E–06	0.00E+00	0.00E+00	3.48E–06	9.06E–03	5.37E-10	2.02E–02	7.44E–04	0	1.81E–03	3.38E–16	7.08E–01
PFC-41-12	9484	a			0.00E+00	1.78E–02	0.00E+00	0.00E+00	1.65E–02	4.30E+01	2.55E–06	9.60E+01	3.53E+00	0	8.59E+00	1.60E–12	3.36E+03
PFC-51-14	8780	a			0.00E+00	1.65E–02	0.00E+00	0.00E+00	1.53E–02	3.98E+01	2.36E–06	8.89E+01	3.26E+00	0	7.95E+00	1.48E–12	3.11E+03
PFC-61-16	8681	a			0.00E+00	1.63E–02	0.00E+00	0.00E+00	1.51E–02	3.93E+01	2.33E–06	8.79E+01	3.23E+00	0	7.86E+00	1.47E–12	3.07E+03
PFC-71-18	8456	a			0.00E+00	1.59E–02	0.00E+00	0.00E+00	1.47E–02	3.83E+01	2.27E–06	8.56E+01	3.14E+00	0	7.66E+00	1.43E–12	2.99E+03
PFC-91-18	7977	a			0.00E+00	1.50E–02	0.00E+00	0.00E+00	1.39E–02	3.61E+01	2.14E–06	8.08E+01	2.97E+00	0	7.23E+00	1.35E–12	2.82E+03
Perfluorodecalin(cis)	8033	a			0.00E+00	1.51E–02	0.00E+00	0.00E+00	1.40E–02	3.64E+01	2.16E–06	8.13E+01	2.99E+00	0	7.28E+00	1.36E–12	2.84E+03

Perfluorodecalin(trans)	6980	a			0.00E+00	1.31E–02	0.00E+00	0.00E+00	1.21E–02	3.16E+01	1.88E–06	7.07E+01	2.60E+00	0	6.32E+00	1.18E–12	2.47E+03
PFC-1114	1	a			0.00E+00	1.88E–06	0.00E+00	0.00E+00	1.74E–06	4.53E–03	2.69E-10	1.01E–02	3.72E–04	0	9.06E–04	1.69E–16	3.54E–01
PFC-1216	1	a			0.00E+00	1.88E–06	0.00E+00	0.00E+00	1.74E–06	4.53E–03	2.69E-10	1.01E–02	3.72E–04	0	9.06E–04	1.69E–16	3.54E–01
Perfluorobuta-1,3-diene	1	a			0.00E+00	1.88E–06	0.00E+00	0.00E+00	1.74E–06	4.53E–03	2.69E-10	1.01E–02	3.72E–04	0	9.06E–04	1.69E–16	3.54E–01
Perfluorobut-1-ene	1	a			0.00E+00	1.88E–06	0.00E+00	0.00E+00	1.74E–06	4.53E–03	2.69E-10	1.01E–02	3.72E–04	0	9.06E–04	1.69E–16	3.54E–01
Perfluorobut-2-ene	2	a			0.00E+00	3.76E–06	0.00E+00	0.00E+00	3.48E–06	9.06E–03	5.37E-10	2.02E–02	7.44E–04	0	1.81E–03	3.38E–16	7.08E–01
HFE-125	13951	a			0.00E+00	2.62E–02	0.00E+00	0.00E+00	2.43E–02	6.32E+01	3.75E–06	1.41E+02	5.19E+00	0	1.26E+01	2.36E–12	4.94E+03
HFE-134 (HG–00)	6512	a			0.00E+00	1.22E–02	0.00E+00	0.00E+00	1.13E–02	2.95E+01	1.75E–06	6.59E+01	2.42E+00	0	5.90E+00	1.10E–12	2.30E+03
HFE-143a	632	a			0.00E+00	1.19E–03	0.00E+00	0.00E+00	1.10E–03	2.86E+00	1.70E–07	6.40E+00	2.35E–01	0	5.72E–01	1.07E–13	2.24E+02
HFE-227ea	7377	a			0.00E+00	1.39E–02	0.00E+00	0.00E+00	1.28E–02	3.34E+01	1.98E–06	7.47E+01	2.74E+00	0	6.68E+00	1.25E–12	2.61E+03
HCFE-235ca2 (enflurane)	705	a			0.00E+00	1.33E–03	0.00E+00	0.00E+00	1.23E–03	3.19E+00	1.89E–07	7.14E+00	2.62E–01	0	6.39E–01	1.19E–13	2.50E+02
HCFE-235da2 (isoflurane)	595	a			0.00E+00	1.12E–03	0.00E+00	0.00E+00	1.04E–03	2.69E+00	1.60E–07	6.02E+00	2.21E–01	0	5.39E–01	1.01E–13	2.11E+02
HFE-236ca	4990	a			0.00E+00	9.38E–03	0.00E+00	0.00E+00	8.68E–03	2.26E+01	1.34E–06	5.05E+01	1.86E+00	0	4.52E+00	8.43E–13	1.77E+03
HFE-236ea2 (desflurane)	2143	a			0.00E+00	4.03E–03	0.00E+00	0.00E+00	3.73E–03	9.71E+00	5.76E–07	2.17E+01	7.97E–01	0	1.94E+00	3.62E–13	7.58E+02
HFE-236fa	1177	a			0.00E+00	2.21E–03	0.00E+00	0.00E+00	2.05E–03	5.33E+00	3.16E–07	1.19E+01	4.38E–01	0	1.07E+00	1.99E–13	4.17E+02
HFE-245cb2	790	a			0.00E+00	1.49E–03	0.00E+00	0.00E+00	1.37E–03	3.58E+00	2.12E–07	8.00E+00	2.94E–01	0	7.16E–01	1.34E–13	2.80E+02
HFE-245fa1	997	a			0.00E+00	1.87E–03	0.00E+00	0.00E+00	1.73E–03	4.52E+00	2.68E–07	1.01E+01	3.71E–01	0	9.03E–01	1.68E–13	3.53E+02
HFE-245fa2	981	a			0.00E+00	1.84E–03	0.00E+00	0.00E+00	1.71E–03	4.44E+00	2.64E–07	9.93E+00	3.65E–01	0	8.89E–01	1.66E–13	3.47E+02
2,2,3,3,3-Penta-fluoropropan-1-ol	23	a			0.00E+00	4.32E–05	0.00E+00	0.00E+00	4.00E–05	1.04E–01	6.18E–09	2.33E–01	8.55E–03	0	2.08E–02	3.89E–15	8.14E+00
HFE-254cb1	365	a			0.00E+00	6.86E–04	0.00E+00	0.00E+00	6.35E–04	1.65E+00	9.81E–08	3.69E+00	1.36E–01	0	3.31E–01	6.17E–14	1.29E+02
HFE-263fb2	2	a			0.00E+00	3.76E–06	0.00E+00	0.00E+00	3.48E–06	9.06E–03	5.37E-10	2.02E–02	7.44E–04	0	1.81E–03	3.38E–16	7.08E–01

Table 8.1 contd. …

...Table 8.1 contd.

Substance	GWP–100	Ref.	ODP	Ref.	YLL, ozone depletion	YLL, climate change	Skin cancer, ozone depletion	Low vision, ozone depletion	Under-nutrition, climate change	Working capacity, climate change	Diarrhea, climate change	Crop, climate change	Meat, climate change	Wood, climate change	Drinking water, climate change	Bio-diversity, climate change	Impact value ($)
HFE-263m1	36	a			0.00E+00	6.77E–05	0.00E+00	0.00E+00	6.26E–05	1.63E–01	9.67E–09	3.64E–01	1.34E–02	0	3.26E–02	6.08E–15	1.27E+01
3,3,3-Trifluoropropan-1-ol	1	a			0.00E+00	1.88E–06	0.00E+00	0.00E+00	1.74E–06	4.53E–03	2.69E-10	1.01E–02	3.72E–04	0	9.06E–04	1.69E–16	3.54E–01
HFE-329mcc2	3598	a			0.00E+00	6.76E–03	0.00E+00	0.00E+00	6.26E–03	1.63E+01	9.67E–07	3.64E+01	1.34E+00	0	3.26E+00	6.08E–13	1.27E+03
HFE-338mmz1	3081	a			0.00E+00	5.79E–03	0.00E+00	0.00E+00	5.36E–03	1.40E+01	8.28E–07	3.12E+01	1.15E+00	0	2.79E+00	5.21E–13	1.09E+03
HFE-338mcf2	1118	a			0.00E+00	2.10E–03	0.00E+00	0.00E+00	1.95E–03	5.06E+00	3.00E–07	1.13E+01	4.16E–01	0	1.01E+00	1.89E–13	3.96E+02
Sevoflurane (HFE-347mmz1)	262	a			0.00E+00	4.93E–04	0.00E+00	0.00E+00	4.56E–04	1.19E+00	7.04E–08	2.65E+00	9.74E–02	0	2.37E–01	4.43E–14	9.27E+01
HFE-347mcc3 (HFE-7000)	641	a			0.00E+00	1.21E–03	0.00E+00	0.00E+00	1.12E–03	2.90E+00	1.72E–07	6.49E+00	2.38E–01	0	5.81E–01	1.08E–13	2.27E+02
HFE-347mcf2	1028	a			0.00E+00	1.93E–03	0.00E+00	0.00E+00	1.79E–03	4.66E+00	2.76E–07	1.04E+01	3.82E–01	0	9.31E–01	1.74E–13	3.64E+02
HFE-347pcf2	1072	a			0.00E+00	2.02E–03	0.00E+00	0.00E+00	1.87E–03	4.86E+00	2.88E–07	1.09E+01	3.99E–01	0	9.71E–01	1.81E–13	3.79E+02
HFE-347mmy1	440	a			0.00E+00	8.27E–04	0.00E+00	0.00E+00	7.66E–04	1.99E+00	1.18E–07	4.45E+00	1.64E–01	0	3.99E–01	7.44E–14	1.56E+02
HFE-356mec3	468	a			0.00E+00	8.80E–04	0.00E+00	0.00E+00	8.14E–04	2.12E+00	1.26E–07	4.74E+00	1.74E–01	0	4.24E–01	7.91E–14	1.66E+02
HFE-356mff2	20	a			0.00E+00	3.76E–05	0.00E+00	0.00E+00	3.48E–05	9.06E–02	5.37E–09	2.02E–01	7.44E–03	0	1.81E–02	3.38E–15	7.08E+00
HFE-356pcf2	867	a			0.00E+00	1.63E–03	0.00E+00	0.00E+00	1.51E–03	3.93E+00	2.33E–07	8.78E+00	3.22E–01	0	7.85E–01	1.47E–13	3.07E+02
HFE-356pcf3	540	a			0.00E+00	1.02E–03	0.00E+00	0.00E+00	9.40E–04	2.45E+00	1.45E–07	5.47E+00	2.01E–01	0	4.89E–01	9.13E–14	1.91E+02
HFE-356pcc3	500	a			0.00E+00	9.40E–04	0.00E+00	0.00E+00	8.70E–04	2.26E+00	1.34E–07	5.06E+00	1.86E–01	0	4.53E–01	8.45E–14	1.77E+02
HFE-356mmz1	17	a			0.00E+00	3.20E–05	0.00E+00	0.00E+00	2.96E–05	7.70E–02	4.57E–09	1.72E–01	6.32E–03	0	1.54E–02	2.87E–15	6.02E+00
HFE-365mcf3	1	a			0.00E+00	1.88E–06	0.00E+00	0.00E+00	1.74E–06	4.53E–03	2.69E-10	1.01E–02	3.72E–04	0	9.06E–04	1.69E–16	3.54E–01
HFE-365mcf2	71	a			0.00E+00	1.33E–04	0.00E+00	0.00E+00	1.24E–04	3.22E–01	1.91E–08	7.19E–01	2.64E–02	0	6.43E–02	1.20E–14	2.51E+01

HFE-374pc2	758	a			0.00E+00	1.43E−03	0.00E+00	0.00E+00	1.32E−03	3.43E+00	2.04E−07	7.67E+00	2.82E−01	0	6.87E−01	1.28E−13	2.68E+02
4,4,4-Trifluorobutan-1-ol	1	a			0.00E+00	1.88E−06	0.00E+00	0.00E+00	1.74E−06	4.53E−03	2.69E-10	1.01E−02	3.72E−04	0	9.06E−04	1.69E−16	3.54E−01
2,2,3,3,4,4,5,5-Octafluorocyclopentanol	16	a			0.00E+00	3.01E−05	0.00E+00	0.00E+00	2.78E−05	7.25E−02	4.30E−09	1.62E−01	5.95E−03	0	1.45E−02	2.70E−15	5.66E+00
HFE-43-10pccc124 (H-Galden 1040x, HG-11)	3353	a			0.00E+00	6.30E−03	0.00E+00	0.00E+00	5.83E−03	1.52E+01	9.01E−07	3.39E+01	1.25E+00	0	3.04E+00	5.67E−13	1.19E+03
HFE-449s1 (HFE-7100)	509	a			0.00E+00	9.57E−04	0.00E+00	0.00E+00	8.86E−04	2.31E+00	1.37E−07	5.15E+00	1.89E−01	0	4.61E−01	8.60E−14	1.80E+02
n-HFE-7100	587	a			0.00E+00	1.10E−03	0.00E+00	0.00E+00	1.02E−03	2.66E+00	1.58E−07	5.94E+00	2.18E−01	0	5.32E−01	9.92E−14	2.08E+02
i-HFE-7100	492	a			0.00E+00	9.25E−04	0.00E+00	0.00E+00	8.56E−04	2.23E+00	1.32E−07	4.98E+00	1.83E−01	0	4.46E−01	8.31E−14	1.74E+02
HFE-569sf2 (HFE-7200)	69	a			0.00E+00	1.30E−04	0.00E+00	0.00E+00	1.20E−04	3.13E−01	1.85E−08	6.98E−01	2.57E−02	0	6.25E−02	1.17E−14	2.44E+01
n-HFE-7200	79	a			0.00E+00	1.49E−04	0.00E+00	0.00E+00	1.37E−04	3.58E−01	2.12E−08	8.00E−01	2.94E−02	0	7.16E−02	1.34E−14	2.80E+01
i-HFE-7200	54	a			0.00E+00	1.02E−04	0.00E+00	0.00E+00	9.40E−05	2.45E−01	1.45E−08	5.47E−01	2.01E−02	0	4.89E−02	9.13E−15	1.91E+01
HFE-236ca12 (HG-10)	6260	a			0.00E+00	1.18E−02	0.00E+00	0.00E+00	1.09E−02	2.84E+01	1.68E−06	6.34E+01	2.33E+00	0	5.67E+00	1.06E−12	2.22E+03
HFE-338pcc13 (HG-01)	3466	a			0.00E+00	6.52E−03	0.00E+00	0.00E+00	6.03E−03	1.57E+01	9.31E−07	3.51E+01	1.29E+00	0	3.14E+00	5.86E−13	1.23E+03
1,1,1,3,3,3-Hexa-fluoropropan-2-ol	221	a			0.00E+00	4.15E−04	0.00E+00	0.00E+00	3.85E−04	1.00E+00	5.94E−08	2.24E+00	8.22E−02	0	2.00E−01	3.73E−14	7.82E+01
HG−02	3250	a			0.00E+00	6.11E−03	0.00E+00	0.00E+00	5.66E−03	1.47E+01	8.73E−07	3.29E+01	1.21E+00	0	2.94E+00	5.49E−13	1.15E+03
HG−03	3400	a			0.00E+00	6.39E−03	0.00E+00	0.00E+00	5.92E−03	1.54E+01	9.14E−07	3.44E+01	1.26E+00	0	3.08E+00	5.75E−13	1.20E+03
HG-20	6201	a			0.00E+00	1.17E−02	0.00E+00	0.00E+00	1.08E−02	2.81E+01	1.67E−06	6.28E+01	2.31E+00	0	5.62E+00	1.05E−12	2.19E+03
HG-21	4628	a			0.00E+00	8.70E−03	0.00E+00	0.00E+00	8.05E−03	2.10E+01	1.24E−06	4.68E+01	1.72E+00	0	4.19E+00	7.82E−13	1.64E+03
HG-30	8575	a			0.00E+00	1.61E−02	0.00E+00	0.00E+00	1.49E−02	3.88E+01	2.30E−06	8.68E+01	3.19E+00	0	7.77E+00	1.45E−12	3.04E+03
1-Ethoxy-1,1,2,2,3,3,3-heptafluoropropane	74	a			0.00E+00	1.39E−04	0.00E+00	0.00E+00	1.29E−04	3.35E−01	1.99E−08	7.49E−01	2.75E−02	0	6.70E−02	1.25E−14	2.62E+01

Table 8.1 contd. …

Table 8.1 contd. …

...Table 8.1 contd.

Substance	GWP–100	Ref.	ODP	Ref.	YLL, ozone depletion	YLL, climate change	Skin cancer, ozone depletion	Low vision, ozone depletion	Under-nutrition, climate change	Working capacity, climate change	Diarrhea, climate change	Crop, climate change	Meat, climate change	Wood, climate change	Drinking water, climate change	Bio-diversity, climate change	Impact value ($)
Fluoroxene	1	a			0.00E+00	1.88E–06	0.00E+00	0.00E+00	1.74E–06	4.53E–03	2.69E-10	1.01E–02	3.72E–04	0	9.06E–04	1.69E–16	3.54E–01
1,1,2,2-Tetrafluoro-1 -(fluoromethoxy)ethane	1051	a			0.00E+00	1.98E–03	0.00E+00	0.00E+00	1.83E–03	4.76E+00	2.82E–07	1.06E+01	3.91E–01	0	9.52E–01	1.78E–13	3.72E+02
2-Ethoxy-3,3,4,4,5-pentafluorotetrahydro-2,5-bis[1,2,2,2-tetrafluoro-1-(trifluoromethyl)ethyl]-furan	68	a			0.00E+00	1.28E–04	0.00E+00	0.00E+00	1.18E–04	3.08E–01	1.83E–08	6.88E–01	2.53E–02	0	6.16E–02	1.15E–14	2.41E+01
Fluoro(methoxy)methane	15	a			0.00E+00	2.82E–05	0.00E+00	0.00E+00	2.61E–05	6.79E–02	4.03E–09	1.52E–01	5.58E–03	0	1.36E–02	2.54E–15	5.31E+00
Difluoro(methoxy)methane	175	a			0.00E+00	3.29E–04	0.00E+00	0.00E+00	3.05E–04	7.93E–01	4.70E–08	1.77E+00	6.51E–02	0	1.59E–01	2.96E–14	6.19E+01
Fluoro(fluoromethoxy) methane	159	a			0.00E+00	2.99E–04	0.00E+00	0.00E+00	2.77E–04	7.20E–01	4.27E–08	1.61E+00	5.91E–02	0	1.44E–01	2.69E–14	5.63E+01
Difluoro(fluoromethoxy) methane	748	a			0.00E+00	1.41E–03	0.00E+00	0.00E+00	1.30E–03	3.39E+00	2.01E–07	7.57E+00	2.78E–01	0	6.78E–01	1.26E–13	2.65E+02
Trifluoro(fluoromethoxy) methane	909	a			0.00E+00	1.71E–03	0.00E+00	0.00E+00	1.58E–03	4.12E+00	2.44E–07	9.20E+00	3.38E–01	0	8.23E–01	1.54E–13	3.22E+02
HG′–01	269	a			0.00E+00	5.06E–04	0.00E+00	0.00E+00	4.68E–04	1.22E+00	7.23E–08	2.72E+00	1.00E–01	0	2.44E–01	4.55E–14	9.52E+01
HG′–02	287	a			0.00E+00	5.40E–04	0.00E+00	0.00E+00	4.99E–04	1.30E+00	7.71E–08	2.91E+00	1.07E–01	0	2.60E–01	4.85E–14	1.02E+02
HG′–03	268	a			0.00E+00	5.04E–04	0.00E+00	0.00E+00	4.66E–04	1.21E+00	7.20E–08	2.71E+00	9.96E–02	0	2.43E–01	4.53E–14	9.49E+01
HFE-329me3	5241	a			0.00E+00	9.85E–03	0.00E+00	0.00E+00	9.12E–03	2.37E+01	1.41E–06	5.31E+01	1.95E+00	0	4.75E+00	8.86E–13	1.85E+03
3,3,4,4, 5,5,6,6,7,7,7-Undecafluoroheptan-1-ol	1	a			0.00E+00	1.88E–06	0.00E+00	0.00E+00	1.74E–06	4.53E–03	2.69E-10	1.01E–02	3.72E–04	0	9.06E–04	1.69E–16	3.54E–01

3,3,4,4,5,5,6,6,7,7,8,8,9,9, 9-Pentadecafluorononan-1-ol	1	a			0.00E+00	1.88E–06	0.00E+00	0.00E+00	1.74E–06	4.53E–03	2.69E-10	1.01E–02	3.72E–04	0	9.06E–04	1.69E–16	3.54E–01
3,3,4,4,5,5,6,6,7,7,8,8,9,9, 10,10, 11,11,11-Nonadecafluoroundecan-1-ol	1	a			0.00E+00	1.88E–06	0.00E+00	0.00E+00	1.74E–06	4.53E–03	2.69E-10	1.01E–02	3.72E–04	0	9.06E–04	1.69E–16	3.54E–01
2-Chloro-1,1,2-trifluoro-1-methoxyethane	149	a			0.00E+00	2.80E–04	0.00E+00	0.00E+00	2.59E–04	6.75E–01	4.00E–08	1.51E+00	5.54E–02	0	1.35E–01	2.52E–14	5.27E+01
PFPMIE (perfluoropolymethyl isopropyl ether)	10789	a			0.00E+00	2.03E–02	0.00E+00	0.00E+00	1.88E–02	4.89E+01	2.90E–06	1.09E+02	4.01E+00	0	9.77E+00	1.82E–12	3.82E+03
HFE-216	1	a			0.00E+00	1.88E–06	0.00E+00	0.00E+00	1.74E–06	4.53E–03	2.69E-10	1.01E–02	3.72E–04	0	9.06E–04	1.69E–16	3.54E–01
Trifluoromethyl formate	712	a			0.00E+00	1.34E–03	0.00E+00	0.00E+00	1.24E–03	3.22E+00	1.91E–07	7.21E+00	2.65E–01	0	6.45E–01	1.20E–13	2.52E+02
Perfluoroethyl formate	703	a			0.00E+00	1.32E–03	0.00E+00	0.00E+00	1.22E–03	3.18E+00	1.89E–07	7.12E+00	2.61E–01	0	6.37E–01	1.19E–13	2.49E+02
Perfluoropropyl formate	456	a			0.00E+00	8.57E–04	0.00E+00	0.00E+00	7.93E–04	2.07E+00	1.23E–07	4.62E+00	1.70E–01	0	4.13E–01	7.71E–14	1.61E+02
Perfluorobutyl formate	475	a			0.00E+00	8.93E–04	0.00E+00	0.00E+00	8.27E–04	2.15E+00	1.28E–07	4.81E+00	1.77E–01	0	4.30E–01	8.03E–14	1.68E+02
2,2,2-Trifluoroethyl formate	41	a			0.00E+00	7.71E–05	0.00E+00	0.00E+00	7.13E–05	1.86E–01	1.10E–08	4.15E–01	1.52E–02	0	3.71E–02	6.93E–15	1.45E+01
3,3,3-Trifluoropropyl formate	21	a			0.00E+00	3.95E–05	0.00E+00	0.00E+00	3.65E–05	9.51E–02	5.64E–09	2.13E–01	7.81E–03	0	1.90E–02	3.55E–15	7.43E+00
1,2,2,2-Tetrafluoroethyl formate	569	a			0.00E+00	1.07E–03	0.00E+00	0.00E+00	9.90E–04	2.58E+00	1.53E–07	5.76E+00	2.12E–01	0	5.15E–01	9.62E–14	2.01E+02
1,1,1,3,3,3-Hexafluoropropan-2-yl formate	403	a			0.00E+00	7.58E–04	0.00E+00	0.00E+00	7.01E–04	1.83E+00	1.08E–07	4.08E+00	1.50E–01	0	3.65E–01	6.81E–14	1.43E+02
Perfluorobutyl acetate	2	a			0.00E+00	3.76E–06	0.00E+00	0.00E+00	3.48E–06	9.06E–03	5.37E-10	2.02E–02	7.44E–04	0	1.81E–03	3.38E–16	7.08E–01
Perfluoropropyl acetate	2	a			0.00E+00	3.76E–06	0.00E+00	0.00E+00	3.48E–06	9.06E–03	5.37E-10	2.02E–02	7.44E–04	0	1.81E–03	3.38E–16	7.08E–01

Table 8.1 contd. …

...Table 8.1 contd.

Substance	GWP–100	Ref.	ODP	Ref.	YLL, ozone depletion	YLL, climate change	Skin cancer, ozone depletion	Low vision, ozone depletion	Under-nutrition, climate change	Working capacity, climate change	Diarrhea, climate change	Crop, climate change	Meat, climate change	Wood, climate change	Drinking water, climate change	Bio-diversity, climate change	Impact value ($)
Perfluoroethyl acetate	3	a			0.00E+00	5.64E–06	0.00E+00	0.00E+00	5.22E–06	1.36E–02	8.06E-10	3.04E–02	1.12E–03	0	2.72E–03	5.07E–16	1.06E+00
Trifluoromethyl acetate	3	a			0.00E+00	5.64E–06	0.00E+00	0.00E+00	5.22E–06	1.36E–02	8.06E-10	3.04E–02	1.12E–03	0	2.72E–03	5.07E–16	1.06E+00
Methyl carbonofluoridate	116	a			0.00E+00	2.18E–04	0.00E+00	0.00E+00	2.02E–04	5.25E–01	3.12E–08	1.17E+00	4.31E–02	0	1.05E–01	1.96E–14	4.11E+01
1,1-Difluoroethyl carbonofluoridate	33	a			0.00E+00	6.20E–05	0.00E+00	0.00E+00	5.74E–05	1.49E–01	8.87E–09	3.34E–01	1.23E–02	0	2.99E–02	5.58E–15	1.17E+01
1,1-Difluoroethyl 2,2,2-trifluoroacetate	38	a			0.00E+00	7.14E–05	0.00E+00	0.00E+00	6.61E–05	1.72E–01	1.02E–08	3.85E–01	1.41E–02	0	3.44E–02	6.42E–15	1.34E+01
Ethyl 2,2,2-trifluoroacetate	2	a			0.00E+00	3.76E–06	0.00E+00	0.00E+00	3.48E–06	9.06E–03	5.37E-10	2.02E–02	7.44E–04	0	1.81E–03	3.38E–16	7.08E–01
2,2,2-Trifluoroethyl 2,2,2-trifluoroacetate	8	a			0.00E+00	1.50E–05	0.00E+00	0.00E+00	1.39E–05	3.62E–02	2.15E–09	8.10E–02	2.97E–03	0	7.25E–03	1.35E–15	2.83E+00
Methyl 2,2,2-trifluoroacetate	64	a			0.00E+00	1.20E–04	0.00E+00	0.00E+00	1.11E–04	2.90E–01	1.72E–08	6.48E–01	2.38E–02	0	5.80E–02	1.08E–14	2.27E+01
Methyl 2,2-difluoroacetate	4	a			0.00E+00	7.52E–06	0.00E+00	0.00E+00	6.96E–06	1.81E–02	1.07E–09	4.05E–02	1.49E–03	0	3.62E–03	6.76E–16	1.42E+00
Difluoromethyl 2,2,2-trifluoroacetate	33	a			0.00E+00	6.20E–05	0.00E+00	0.00E+00	5.74E–05	1.49E–01	8.87E–09	3.34E–01	1.23E–02	0	2.99E–02	5.58E–15	1.17E+01
2,2,3,3,4,4,4-Hepta-fluorobutan-1-ol	41	a			0.00E+00	7.71E–05	0.00E+00	0.00E+00	7.13E–05	1.86E–01	1.10E–08	4.15E–01	1.52E–02	0	3.71E–02	6.93E–15	1.45E+01
1,1,2-Trifluoro-2-(trifluoromethoxy)-ethane	1489	a			0.00E+00	2.80E–03	0.00E+00	0.00E+00	2.59E–03	6.74E+00	4.00E–07	1.51E+01	5.54E–01	0	1.35E+00	2.52E–13	5.27E+02
1-Ethoxy-1,1,2,3,3,3-hexafluoropropane	28	a			0.00E+00	5.26E–05	0.00E+00	0.00E+00	4.87E–05	1.27E–01	7.52E–09	2.83E–01	1.04E–02	0	2.54E–02	4.73E–15	9.91E+00

1,1,1,2,2,3,3-Heptafluoro-3-(1,2,2,2-tetrafluoroethoxy)-propane	7371	a			0.00E+00	1.39E–02	0.00E+00	0.00E+00	1.28E–02	3.34E+01	1.98E–06	7.46E+01	2.74E+00	0	6.68E+00	1.25E–12	2.61E+03
2,2,3,3-Tetrafluoro-1-propanol	16	a			0.00E+00	3.01E–05	0.00E+00	0.00E+00	2.78E–05	7.25E–02	4.30E–09	1.62E–01	5.95E–03	0	1.45E–02	2.70E–15	5.66E+00
2,2,3,4,4,4-Hexafluoro-1-butanol Hexafluoropropyl methyl ether	21	a			0.00E+00	3.95E–05	0.00E+00	0.00E+00	3.65E–05	9.51E–02	5.64E–09	2.13E–01	7.81E–03	0	1.90E–02	3.55E–15	7.43E+00
2,2,3,3,4,4,4-Heptafluoro-1-butanol	20	a			0.00E+00	3.76E–05	0.00E+00	0.00E+00	3.48E–05	9.06E–02	5.37E–09	2.02E–01	7.44E–03	0	1.81E–02	3.38E–15	7.08E+00
1,1,2,2-Tetrafluoro-3-methoxy-propane	1	a			0.00E+00	1.88E–06	0.00E+00	0.00E+00	1.74E–06	4.53E–03	2.69E-10	1.01E–02	3.72E–04	0	9.06E–04	1.69E–16	3.54E–01
Perfluoro-2-methyl-3-pentanone	1	a			0.00E+00	1.88E–06	0.00E+00	0.00E+00	1.74E–06	4.53E–03	2.69E-10	1.01E–02	3.72E–04	0	9.06E–04	1.69E–16	3.54E–01
3,3,3-Trifluoro-propanal	1	a			0.00E+00	1.88E–06	0.00E+00	0.00E+00	1.74E–06	4.53E–03	2.69E-10	1.01E–02	3.72E–04	0	9.06E–04	1.69E–16	3.54E–01
2-Fluoroethanol	1	a			0.00E+00	1.88E–06	0.00E+00	0.00E+00	1.74E–06	4.53E–03	2.69E-10	1.01E–02	3.72E–04	0	9.06E–04	1.69E–16	3.54E–01
2,2-Difluoroethanol	4	a			0.00E+00	7.52E–06	0.00E+00	0.00E+00	6.96E–06	1.81E–02	1.07E–09	4.05E–02	1.49E–03	0	3.62E–03	6.76E–16	1.42E+00
2,2,2-Trifluoroethanol	24	a			0.00E+00	4.51E–05	0.00E+00	0.00E+00	4.18E–05	1.09E–01	6.45E–09	2.43E–01	8.92E–03	0	2.17E–02	4.06E–15	8.49E+00
1,1′-Oxybis[2-(difluoromethoxy) -1,1,2,2-tetrafluoroethane	5741	a			0.00E+00	1.08E–02	0.00E+00	0.00E+00	9.99E–03	2.60E+01	1.54E–06	5.81E+01	2.13E+00	0	5.20E+00	9.70E–13	2.03E+03
1,1,3,3,4,4,6,6,7,7,9,9, 10,10,12,12-hexadecafluoro-2,5,8,11-Tetraoxadodecane	5245	a			0.00E+00	9.86E–03	0.00E+00	0.00E+00	9.13E–03	2.38E+01	1.41E–06	5.31E+01	1.95E+00	0	4.75E+00	8.86E–13	1.86E+03

Table 8.1 contd. …

...Table 8.1 contd.

Substance	GWP–100	Ref.	ODP	Ref.	YLL, ozone depletion	YLL, climate change	Skin cancer, ozone depletion	Low vision, ozone depletion	Under-nutrition, climate change	Working capacity, climate change	Diarrhea, climate change	Crop, climate change	Meat, climate change	Wood, climate change	Drinking water, climate change	Bio-diversity, climate change	Impact value ($)
1,1,3,3,4,4,6,6,7,7,9,9,10,10,12,12,13,13,15,15-Eicosafluoro-2,5,8,11,14-Pentaoxapentadecane	4240	a			0.00E+00	7.97E–03	0.00E+00	0.00E+00	7.38E–03	1.92E+01	1.14E–06	4.29E+01	1.58E+00	0	3.84E+00	7.17E–13	1.50E+03
CF4	6300	d			0.00E+00	1.18E–02	0.00E+00	0.00E+00	1.10E–02	2.85E+01	1.69E–06	6.38E+01	2.34E+00	0	5.71E+00	1.06E–12	2.23E+03
C2F6	12500	d			0.00E+00	2.35E–02	0.00E+00	0.00E+00	2.18E–02	5.66E+01	3.36E–06	1.27E+02	4.65E+00	0	1.13E+01	2.11E–12	4.42E+03
c-C4F8	9100	d			0.00E+00	1.71E–02	0.00E+00	0.00E+00	1.58E–02	4.12E+01	2.45E–06	9.21E+01	3.38E+00	0	8.24E+00	1.54E–12	3.22E+03
C6F14	6800	d			0.00E+00	1.28E–02	0.00E+00	0.00E+00	1.18E–02	3.08E+01	1.83E–06	6.88E+01	2.53E+00	0	6.16E+00	1.15E–12	2.41E+03
CHFBr2	100	e	1	h	1.05E–04	1.88E–04	1.33E–06	2.66E–04	1.74E–04	4.53E–01	2.69E–08	1.01E+00	3.72E–02	0	9.06E–02	1.69E–14	5.15E+01
C2HFBr4	100		0.5	h	5.25E–05	1.88E–04	6.65E–07	1.33E–04	1.74E–04	4.53E–01	2.69E–08	1.01E+00	3.72E–02	0	9.06E–02	1.69E–14	4.34E+01
C2HF2Br3	100		1.1	h	1.16E–04	1.88E–04	1.46E–06	2.93E–04	1.74E–04	4.53E–01	2.69E–08	1.01E+00	3.72E–02	0	9.06E–02	1.69E–14	5.31E+01
C2HF3Br2	100		1	h	1.05E–04	1.88E–04	1.33E–06	2.66E–04	1.74E–04	4.53E–01	2.69E–08	1.01E+00	3.72E–02	0	9.06E–02	1.69E–14	5.15E+01
C2HF4Br	100		0.9	h	9.45E–05	1.88E–04	1.20E–06	2.39E–04	1.74E–04	4.53E–01	2.69E–08	1.01E+00	3.72E–02	0	9.06E–02	1.69E–14	4.99E+01
C2H2FBr3	100		0.5	h	5.25E–05	1.88E–04	6.65E–07	1.33E–04	1.74E–04	4.53E–01	2.69E–08	1.01E+00	3.72E–02	0	9.06E–02	1.69E–14	4.34E+01
C2H2F2Br2	100		0.8	h	8.40E–05	1.88E–04	1.06E–06	2.13E–04	1.74E–04	4.53E–01	2.69E–08	1.01E+00	3.72E–02	0	9.06E–02	1.69E–14	4.83E+01
C2H2F3Br	100		1.1	h	1.16E–04	1.88E–04	1.46E–06	2.93E–04	1.74E–04	4.53E–01	2.69E–08	1.01E+00	3.72E–02	0	9.06E–02	1.69E–14	5.31E+01
C2H3FBr2	100		0.8	h	8.40E–05	1.88E–04	1.06E–06	2.13E–04	1.74E–04	4.53E–01	2.69E–08	1.01E+00	3.72E–02	0	9.06E–02	1.69E–14	4.83E+01
C2H3F2Br	100		0.6	h	6.30E–05	1.88E–04	7.98E–07	1.60E–04	1.74E–04	4.53E–01	2.69E–08	1.01E+00	3.72E–02	0	9.06E–02	1.69E–14	4.50E+01
C2H4FBr	100		0.08	h	8.40E–06	1.88E–04	1.06E–07	2.13E–05	1.74E–04	4.53E–01	2.69E–08	1.01E+00	3.72E–02	0	9.06E–02	1.69E–14	3.67E+01
C3HFBr6	100		0.9	h	9.45E–05	1.88E–04	1.20E–06	2.39E–04	1.74E–04	4.53E–01	2.69E–08	1.01E+00	3.72E–02	0	9.06E–02	1.69E–14	4.99E+01
C3HF2Br5	100		1	h	1.05E–04	1.88E–04	1.33E–06	2.66E–04	1.74E–04	4.53E–01	2.69E–08	1.01E+00	3.72E–02	0	9.06E–02	1.69E–14	5.15E+01
C3HF3Br4	100		1	h	1.05E–04	1.88E–04	1.33E–06	2.66E–04	1.74E–04	4.53E–01	2.69E–08	1.01E+00	3.72E–02	0	9.06E–02	1.69E–14	5.15E+01

C3HF4Br3	100		1.3	h	1.37E–04	1.88E–04	1.73E–06	3.46E–04	1.74E–04	4.53E–01	2.69E–08	1.01E+00	3.72E–02	0	9.06E–02	1.69E–14	5.63E+01
C3HF5Br2	100		1.4	h	1.47E–04	1.88E–04	1.86E–06	3.72E–04	1.74E–04	4.53E–01	2.69E–08	1.01E+00	3.72E–02	0	9.06E–02	1.69E–14	5.79E+01
C3HF6Br	100		2	h	2.10E–04	1.88E–04	2.66E–06	5.32E–04	1.74E–04	4.53E–01	2.69E–08	1.01E+00	3.72E–02	0	9.06E–02	1.69E–14	6.76E+01
C3H2FBr5	100		1	h	1.05E–04	1.88E–04	1.33E–06	2.66E–04	1.74E–04	4.53E–01	2.69E–08	1.01E+00	3.72E–02	0	9.06E–02	1.69E–14	5.15E+01
C3H2F2Br4	100		1.1	h	1.16E–04	1.88E–04	1.46E–06	2.93E–04	1.74E–04	4.53E–01	2.69E–08	1.01E+00	3.72E–02	0	9.06E–02	1.69E–14	5.31E+01
C3H2F3Br3	100		2.9	h	3.05E–04	1.88E–04	3.86E–06	7.71E–04	1.74E–04	4.53E–01	2.69E–08	1.01E+00	3.72E–02	0	9.06E–02	1.69E–14	8.21E+01
C3H2F4Br2	100		3.9	h	4.10E–04	1.88E–04	5.19E–06	1.04E–03	1.74E–04	4.53E–01	2.69E–08	1.01E+00	3.72E–02	0	9.06E–02	1.69E–14	9.81E+01
C3H2F5Br	100		7.4	h	7.77E–04	1.88E–04	9.84E–06	1.97E–03	1.74E–04	4.53E–01	2.69E–08	1.01E+00	3.72E–02	0	9.06E–02	1.69E–14	1.54E+02
C3H3FBr4	100		1.3	h	1.37E–04	1.88E–04	1.73E–06	3.46E–04	1.74E–04	4.53E–01	2.69E–08	1.01E+00	3.72E–02	0	9.06E–02	1.69E–14	5.63E+01
C3H3F2Br3	100		1.6	h	1.68E–04	1.88E–04	2.13E–06	4.26E–04	1.74E–04	4.53E–01	2.69E–08	1.01E+00	3.72E–02	0	9.06E–02	1.69E–14	6.11E+01
C3H3F3Br2	100		1.2	h	1.26E–04	1.88E–04	1.60E–06	3.19E–04	1.74E–04	4.53E–01	2.69E–08	1.01E+00	3.72E–02	0	9.06E–02	1.69E–14	5.47E+01
C3H3F4Br	100		2.3	h	2.42E–04	1.88E–04	3.06E–06	6.12E–04	1.74E–04	4.53E–01	2.69E–08	1.01E+00	3.72E–02	0	9.06E–02	1.69E–14	7.24E+01
C3H4FBr3	100		0.16	h	1.68E–05	1.88E–04	2.13E–07	4.26E–05	1.74E–04	4.53E–01	2.69E–08	1.01E+00	3.72E–02	0	9.06E–02	1.69E–14	3.80E+01
C3H4F2Br2	100		0.5	h	5.25E–05	1.88E–04	6.65E–07	1.33E–04	1.74E–04	4.53E–01	2.69E–08	1.01E+00	3.72E–02	0	9.06E–02	1.69E–14	4.34E+01
C3H4F3Br	100		0.43	h	4.52E–05	1.88E–04	5.72E–07	1.14E–04	1.74E–04	4.53E–01	2.69E–08	1.01E+00	3.72E–02	0	9.06E–02	1.69E–14	4.23E+01
C3H5FBr2	100		0.21	h	2.21E–05	1.88E–04	2.79E–07	5.59E–05	1.74E–04	4.53E–01	2.69E–08	1.01E+00	3.72E–02	0	9.06E–02	1.69E–14	3.88E+01
C3H5F2Br	100		0.43	h	4.52E–05	1.88E–04	5.72E–07	1.14E–04	1.74E–04	4.53E–01	2.69E–08	1.01E+00	3.72E–02	0	9.06E–02	1.69E–14	4.23E+01
C3H6FBr	100		0.36	h	3.78E–05	1.88E–04	4.79E–07	9.58E–05	1.74E–04	4.53E–01	2.69E–08	1.01E+00	3.72E–02	0	9.06E–02	1.69E–14	4.12E+01
CH2BrCl	100		0.12	h	1.26E–05	1.88E–04	1.60E–07	3.19E–05	1.74E–04	4.53E–01	2.69E–08	1.01E+00	3.72E–02	0	9.06E–02	1.69E–14	3.73E+01

an estimated magnitude assuming similarity with H,F and Br containing halo-organic compounds. The ODP values are from UNEP [7] (h), WMO [8] (f), from USEPA (g) and WMO (i).

Uncertainties vary between a factor of 1.2 and 2.5 in GWP and ODP factors. The uncertainty varies most for short lived ozone depletion substances, but they do not influence the monetary impact values more than marginally. The climate impacts undernutrition and decreased working capacity is contributing most to the monetary impact values. An uncertainty of a factor of 2 seems reasonable.

References

1. WHO. *Stratospheric ozone depletion, ultraviolet radiation and health*. Climate change and human health, [cited 2019 March 27]; Available from: http://www.who.int/globalchange/climate/summary/en/index7.html.
2. Consulting, I., *Human Health Benefits of Straospheric Ozone Protection*. 2006, U.S. Environmental Protection Agency: Washington, DC 20460.
3. Burnet, N.G. et al., Years of life lost (YLL) from cancer is an important measure of population burden—and should be considered when allocating research funds. *British Journal of Cancer*, 2005. **92**(2): 241–245.
4. Brustugun, O.T., B. Møller and Å. Helland, Years of life lost as a measure of cancer burden on a national level. *Br. J. Cancer*, 2014. **111**(5): 1014–1020.
5. Ravishankara, A.R., J.S. Daniel and R.W. Portmann, Nitrous Oxide (N₂O): The Dominant Ozone-Depleting Substance Emitted in the 21st Century. *Science*, 2009. **326**(5949): 123–125.
6. Houghton, J.T. et al., *Climate Change 1994: Radiative Forcing of Climate Change and An Evaluation of the IPCC IS92 Emission Scenarios*. 1994, IPCC: Cambridge, England.
7. UNEP, *The Montreal Protocol on Substances that Deplete the Ozone Layer*. 2000, Ozone Secretariat, United Nations Environment Programme: Nairobi, Kenya.
8. (WMO), W.M.O., *GORMP, 50. Scientific Assessment of Ozone Depletion: 2006 : report of the Montreal Protocol Scientific Assessment Panel*. 2007, WMO; National Oceanic and Atmospheric Administration (NOAA); United Nations Environment Programme (UNEP).

9

Emission of Volatile Organic Compounds (VOC) to Air

VOC includes methane. If methane is excluded, the category is called NMVOC, non-methane volatile organic compounds. Most of the NMVOC in the atmosphere originates from vegetation, but human activities like agriculture and silviculture may influence the amount emitted. Anthropogenic emissions of NMVOC is 130 Tg C/yr. About 1350 Tg C as VOC is emitted to air globally each year. Of that 1350 Tg C, 400 Tg C is oxidized in air to CO and CO_2, 800 Tg C is deposited to ground as vapor and 150 TgC is transformed to particles [1]. Global anthropogenic methane emissions is about 300 Tg/yr. Anthropogenic emissions account for 50 to 65% of total methane emissions (IPCC AR5 WG1 TS.2.8.3).

Some VOCs are emitted in large quantities and enough information is available for risk assessments. Others are present in trace amounts and various assumptions must be made on their risks.

9.1 Specification of the Emission

Volatile organic compounds are gaseous in the atmosphere but may be liquid as emissions. For instance, when handling organic solvents, solvent spray may be formed and not evaporated due to high gas concentrations. Later when the spray is diluted in air, the remaining solvent will evaporate.

The emissions valued here are emissions of specific VOCs anywhere in the world in 2015.

9.2 Impact Pathways

There are four mechanisms by which VOC impacts on environmental goods:

1. Direct exposure: Some VOCs are carcinogenic
2. Particle formation: Several VOCs form secondary aerosols when oxidized in air. Secondary organic aerosols (SOA) impact on human health and on climate. The impacts are significant, but complex and uncertain as to their extent. Isoprene, terpenes, sesquiterpenes, alcohols and aromatics are known to form SOA, and quantitative estimations on how much SOA that is formed exist for different types of compounds [2].
3. Formation of tropospheric ozone: VOC interact with NO_x and various radicals formed in sunlight and cause increased ozone concentrations in air. The potential to create ozone is expressed quantitatively in POCP, photochemical ozone creation potential, where ethene is a reference, with the POCP of 1. There are several estimations of POCP values, which varies dependent on local conditions and what type of ozone value that is assessed: average values (over certain number of days), maximum values or average above a certain level [3]. Ideally, different POCP values should be used for impacts on different environmental goods. Average values are most relevant for impacts on materials, maximum daily values are relevant for short term health impacts, averages above 40 $\mu g/m^3$ ($POCP_{AOT40}$) are relevant for crop damage and human health, while averages above 60 and 90 $\mu g/m^3$ are most relevant for human health impacts. The $POCP_{AOT40}$ is here preferred, and differences due to threshold values and locations are dealt with as uncertainties.
4. Climate impacts: The extent of the impacts varies due to local and regional conditions. The residence times in air varies from several years (methane) to a week.

9.3 Environmental Impact Indicators

9.3.1 YLL via climate change

The added environmental impact factors for YLL from CO_2 was 1.88E–06 YLL/kg CO_2 (Sects. 6.1.3.1–6.1.3.4). If GWP for a VOC is GWP_i, the environmental impact factor for the ith compound is 1.88E–06*GWP_i. Environmental impact factors for different VOC calculated in this way are shown in Table 9.2.

The uncertainty in the environmental impact factor for YLL from CO_2 was estimated to 2 in Sec. 6.1.3.2. An additional uncertainty due to varying local and regional conditions of a factor of 2 is assumed, resulting in a total uncertainty of $\exp \sqrt{(\ln 2)^2 + (\ln 2)^2} = 2.7$.

9.3.2 YLL via oxidants

The environmental impact factor for YLL from NO_x via tropospheric ozone was determined to 6.97E–06 YLL/kg NO_x in Sec. 6.3.3.3. The POCP used for NO_x was 0.46. If the POCP for the ith compound is denoted $POCP_i$, then the environmental impact factor for the ith VOC is 6.97E–06/0.46*$POCP_i$. Environmental impact factors for different VOC calculated in this way are shown in Table 9.2.

The uncertainty in the environmental impact factor for YLL from NO_x was estimated to be 2 in Sec. 6.3.3.3. An additional uncertainty due to varying local and regional conditions of a factor of 2 is assumed, resulting in a total uncertainty of $\exp \sqrt{(\ln 2)^2 + (\ln 2)^2} = 2.7$.

9.3.3 YLL via secondary particles

The environmental impact factor was determined in Sec. 7.1.3.1 to be 1.34E–03 YLL/kg PM2.5.

The extent of particle formation from VOC emissions have been estimated by Henze et al. [2] (Table 9.1).

The average particle formation efficiency for all VOC is estimated to be 11% [1].

The environmental impact factor for YLL via secondary particles is 1.34E–03*P YLL/kg VOC measured as carbon, where P is the particle formation efficiency. Environmental impact factors for different VOC:s are shown in Table 9.2.

Table 9.1. Part of VOC emitted to air that is converted to PM2.5.

Substance category	Particle formation efficiency
terpenes	7.2%
alcohols	4.2%
sesquiterpenes	14.2%
isoprene	3.2%
aromatics	18.6%

Table 9.2. Environmental impact factors and monetary values for individual VOCs. References to GWP100, IPPC AR5 WG1 Chapter 8 (a). GWP100 for NMVOC is 4.5 with respect toits carbon content. The GWP100 in the table is recalculated to the individual compound mass. POCPs are from Altensted and Pleijel [12] (b) and Derwent et al. [13] (c). Particle formation potentials are from Henze et al. [2] (d).

VOC compound	GWP 100	Ref	POCP	Ref	Particle formation potential	Ref	YLL via climate change	YLL via oxidants	YLL via secondary particles	YLL via cancer	Under-nutrition via climate change	Working capacity via climate change	Diarrhea via climate change	YLD via secondary particles	Crop loss via climate change	Crop loss via oxidants	Meat prod. capacity via climate change	Fish prod. capacity via climate change	Drinking water via climate change	Biodiversity via climate change	Impact value ($/kg)
Alkanes																					
Methane	28.5	a	0.008	b	0		5.36E–05	1.21E–07	0.00E+00		4.91E–05	1.29E–01	7.67E–09	0.00E+00	4.05E–01	2.09E–03	1.06E–02	1.46E–05	2.58E–02	4.82E-15	10.13
Ethan	3.9	a	0.25	b	0		7.25E–06	3.79E–06	0.00E+00		6.65E–06	1.75E–02	1.04E–09	0.00E+00	5.48E–02	6.52E–02	1.43E–03	0.00E+00	3.49E–03	6.52E-16	1.79
Propane	3.9	a	0.55	b	0		7.25E–06	8.33E–06	0.00E+00		6.65E–06	1.75E–02	1.04E–09	0.00E+00	5.48E–02	1.43E–01	1.43E–03	0.00E+00	3.49E–03	6.52E-16	2.30
N-butane	3.9	a	0.72	b	0		7.25E–06	1.09E–05	0.00E+00		6.65E–06	1.75E–02	1.04E–09	0.00E+00	5.48E–02	1.88E–01	1.43E–03	0.00E+00	3.49E–03	6.52E-16	2.59
I-butane	3.9	a	0.58	b	0		7.25E–06	8.79E–06	0.00E+00		6.65E–06	1.75E–02	1.04E–09	0.00E+00	5.48E–02	1.51E–01	1.43E–03	0.00E+00	3.49E–03	6.52E-16	2.35
N-pentane	3.9	a	0.94	b	0		7.25E–06	1.42E–05	0.00E+00		6.65E–06	1.75E–02	1.04E–09	0.00E+00	5.48E–02	2.45E–01	1.43E–03	0.00E+00	3.49E–03	6.52E-16	2.97
I-pentane	3.9	a	0.45	b	0		7.25E–06	6.82E–06	0.00E+00		6.65E–06	1.75E–02	1.04E–09	0.00E+00	5.48E–02	1.17E–01	1.43E–03	0.00E+00	3.49E–03	6.52E-16	2.13
Neopentane	3.9	a	0.173	c	0		7.25E–06	2.62E–06	0.00E+00		6.65E–06	1.75E–02	1.04E–09	0.00E+00	5.48E–02	4.51E–02	1.43E–03	0.00E+00	3.49E–03	6.52E-16	1.66
Hexane	3.9	a	1.04	b	0		7.25E–06	1.58E–05	0.00E+00		6.65E–06	1.75E–02	1.04E–09	0.00E+00	5.48E–02	2.71E–01	1.43E–03	0.00E+00	3.49E–03	6.52E-16	3.13
2-metyl-pentane	3.9	a	0.82	b	0		7.25E–06	1.24E–05	0.00E+00		6.65E–06	1.75E–02	1.04E–09	0.00E+00	5.48E–02	2.14E–01	1.43E–03	0.00E+00	3.49E–03	6.52E-16	2.76
3-metyl-pentane	3.9	a	0.9	b	0		7.25E–06	1.36E–05	0.00E+00		6.65E–06	1.75E–02	1.04E–09	0.00E+00	5.48E–02	2.35E–01	1.43E–03	0.00E+00	3.49E–03	6.52E-16	2.90
2,2-Dime-thylbutane	3.9	a	0.479	b	0		7.25E–06	7.26E–06	0.00E+00		6.65E–06	1.75E–02	1.04E–09	0.00E+00	5.48E–02	1.25E–01	1.43E–03	0.00E+00	3.49E–03	6.52E-16	2.18
2,3-Dime-thylbutane	3.9	a	0.241	c	0		7.25E–06	3.65E–06	0.00E+00		6.65E–06	1.75E–02	1.04E–09	0.00E+00	5.48E–02	6.29E–02	1.43E–03	0.00E+00	3.49E–03	6.52E-16	1.78
N-heptane	3.9	a	1	b	0		7.25E–06	1.52E–05	0.00E+00		6.65E–06	1.75E–02	1.04E–09	0.00E+00	5.48E–02	2.61E–01	1.43E–03	0.00E+00	3.49E–03	6.52E-16	3.07
2-Me-thylhexane	3.9	a	0.411	c	0		7.25E–06	6.23E–06	0.00E+00		6.65E–06	1.75E–02	1.04E–09	0.00E+00	5.48E–02	1.07E–01	1.43E–03	0.00E+00	3.49E–03	6.52E-16	2.07
3-Me-thylhexane	3.9	a	0.364	c	0		7.25E–06	5.52E–06	0.00E+00		6.65E–06	1.75E–02	1.04E–09	0.00E+00	5.48E–02	9.50E–02	1.43E–03	0.00E+00	3.49E–03	6.52E-16	1.99

N-oktane	3.9	a	0.98	b	0		7.25E–06	1.48E–05	0.00E+00		6.65E–06	1.75E–02	1.04E–09	0.00E+00	5.48E–02	2.56E–01	1.43E–03	0.00E+00	3.49E–03	6.52E-16	3.03
2-metylheptane	3.9	a	0.87	b	0		7.25E–06	1.32E–05	0.00E+00		6.65E–06	1.75E–02	1.04E–09	0.00E+00	5.48E–02	2.27E–01	1.43E–03	0.00E+00	3.49E–03	6.52E-16	2.85
N-nonane	3.9	a	0.97	b	0		7.25E–06	1.47E–05	0.00E+00		6.65E–06	1.75E–02	1.04E–09	0.00E+00	5.48E–02	2.53E–01	1.43E–03	0.00E+00	3.49E–03	6.52E-16	3.02
2-metyloktane	3.9	a	0.87	b	0		7.25E–06	1.32E–05	0.00E+00		6.65E–06	1.75E–02	1.04E–09	0.00E+00	5.48E–02	2.27E–01	1.43E–03	0.00E+00	3.49E–03	6.52E-16	2.85
N-decane	3.9	a	0.95	b	0		7.25E–06	1.44E–05	0.00E+00		6.65E–06	1.75E–02	1.04E–09	0.00E+00	5.48E–02	2.48E–01	1.43E–03	0.00E+00	3.49E–03	6.52E-16	2.98
2-me-thylnonane	3.9	a	0.88	b	0		7.25E–06	1.33E–05	0.00E+00		6.65E–06	1.75E–02	1.04E–09	0.00E+00	5.48E–02	2.30E–01	1.43E–03	0.00E+00	3.49E–03	6.52E-16	2.86
N-undecane	3.9	a	0.92	b	0		7.25E–06	1.39E–05	0.00E+00		6.65E–06	1.75E–02	1.04E–09	0.00E+00	5.48E–02	2.40E–01	1.43E–03	0.00E+00	3.49E–03	6.52E-16	2.93
N-dodecane	3.9	a	0.92	b	0		7.25E–06	1.39E–05	0.00E+00		6.65E–06	1.75E–02	1.04E–09	0.00E+00	5.48E–02	2.40E–01	1.43E–03	0.00E+00	3.49E–03	6.52E-16	2.93
Cycloalkanes																					
Cyclohexane	3.9	a	0.29	c	0		7.25E–06	4.39E–06	0.00E+00		6.65E–06	1.75E–02	1.04E–09	0.00E+00	5.48E–02	7.57E–02	1.43E–03	0.00E+00	3.49E–03	6.52E-16	1.86
Metyl-cyclohexane	3.9	a	0.75	b	0		7.25E–06	1.14E–05	0.00E+00		6.65E–06	1.75E–02	1.04E–09	0.00E+00	5.48E–02	1.96E–01	1.43E–03	0.00E+00	3.49E–03	6.52E-16	2.64
Cyclohexanone	3.3	a	0.299	c	0		6.22E–06	4.53E–06	0.00E+00		5.70E–06	1.50E–02	8.89E-10	0.00E+00	4.69E–02	7.80E–02	1.23E–03	0.00E+00	3.00E–03	5.59E-16	1.68
Cyclohexanol	3.3	a	0.446	c	0		6.22E–06	6.76E–06	0.00E+00		5.70E–06	1.50E–02	8.89E-10	0.00E+00	4.69E–02	1.16E–01	1.23E–03	0.00E+00	3.00E–03	5.59E-16	1.93
Alkenes		a																			
Ethene	3.9	a	1	b,c	0		7.25E–06	1.52E–05	0.00E+00	2.58E–06	6.65E–06	1.75E–02	1.04E–09	0.00E+00	5.48E–02	2.61E–01	1.43E–03	0.00E+00	3.49E–03	6.52E-16	3.34
Propene	3.9	a	1.25	b	0		7.25E–06	1.89E–05	0.00E+00	1.92E–07	6.65E–06	1.75E–02	1.04E–09	0.00E+00	5.48E–02	3.26E–01	1.43E–03	0.00E+00	3.49E–03	6.52E-16	3.51
1-butene	3.9	a	1.26	b	0		7.25E–06	1.91E–05	0.00E+00		6.65E–06	1.75E–02	1.04E–09	0.00E+00	5.48E–02	3.29E–01	1.43E–03	0.00E+00	3.49E–03	6.52E-16	3.51
2-butene	3.9	a	1.13	b	0		7.25E–06	1.71E–05	0.00E+00		6.65E–06	1.75E–02	1.04E–09	0.00E+00	5.48E–02	2.95E–01	1.43E–03	0.00E+00	3.49E–03	6.52E-16	3.29
Methylpropene	3.9	a	0.627	c	0		7.25E–06	9.50E–06	0.00E+00		6.65E–06	1.75E–02	1.04E–09	0.00E+00	5.48E–02	1.64E–01	1.43E–03	0.00E+00	3.49E–03	6.52E-16	2.43
1-pentene	3.9	a	1.21	b	0		7.25E–06	1.83E–05	0.00E+00		6.65E–06	1.75E–02	1.04E–09	0.00E+00	5.48E–02	3.16E–01	1.43E–03	0.00E+00	3.49E–03	6.52E-16	3.42
2-pentene	3.9	a	1.28	b	0		7.25E–06	1.94E–05	0.00E+00		6.65E–06	1.75E–02	1.04E–09	0.00E+00	5.48E–02	3.34E–01	1.43E–03	0.00E+00	3.49E–03	6.52E-16	3.54
2-Methylbutan-1-ol	3.2	a	0.97	b	0		6.04E–06	1.47E–05	0.00E+00		5.54E–06	1.46E–02	8.65E-10	0.00E+00	4.56E–02	2.53E–01	1.20E–03	0.00E+00	2.91E–03	5.43E-16	2.79

Table 9.2 contd. ...

... Table 9.2 contd.

VOC compound	GWP 100	Ref	POCP	Ref	Particle formation potential	Ref	YLL via climate change	YLL via oxidants	YLL via secondary particles	YLL via cancer	Under-nutrition via climate change	Working capacity via climate change	Diarrhea via climate change	YLD via secondary particles	Crop loss via climate change	Crop loss via oxidants	Meat prod. capacity via climate change	Fish prod. capacity via climate change	Drinking water via climate change	Biodiversity via climate change	Impact value ($/kg)
2-methyl-2-butene	3.9	a	1.11	b	0		7.25E–06	1.68E–05	0.00E+00		6.65E–06	1.75E–02	1.04E–09	0.00E+00	5.48E–02	2.90E–01	1.43E–03	0.00E+00	3.49E–03	6.52E-16	3.25
3-Methylbut-1-ene	3.9	a	0.671	b	0		7.25E–06	1.02E–05	0.00E+00		6.65E–06	1.75E–02	1.04E–09	0.00E+00	5.48E–02	1.75E–01	1.43E–03	0.00E+00	3.49E–03	6.52E-16	2.51
Hex-1-ene	3.9	a	0.874	c	0		7.25E–06	1.32E–05	0.00E+00		6.65E–06	1.75E–02	1.04E–09	0.00E+00	5.48E–02	2.28E–01	1.43E–03	0.00E+00	3.49E–03	6.52E-16	2.85
Hex-2-ene	3.9	a	1.07	c	0		7.25E–06	1.62E–05	0.00E+00		6.65E–06	1.75E–02	1.04E–09	0.00E+00	5.48E–02	2.79E–01	1.43E–03	0.00E+00	3.49E–03	6.52E-16	3.19
1,3-Butadiene	3.9	a	0.851	c	0		7.25E–06	1.29E–05	0.00E+00	1.54E–05	6.65E–06	1.75E–02	1.04E–09	0.00E+00	5.48E–02	2.22E–01	1.43E–03	0.00E+00	3.49E–03	6.52E-16	4.46
Dialkenes																					
Methacrolein	3.9	a	1.51	b	0		7.25E–06	2.29E–05	0.00E+00		6.65E–06	1.75E–02	1.04E–09	0.00E+00	5.48E–02	3.94E–01	1.43E–03	0.00E+00	3.49E–03	6.52E-16	3.93
Styrene	3.9	a	0.88	b	0		7.25E–06	1.33E–05	0.00E+00		6.65E–06	1.75E–02	1.04E–09	0.00E+00	5.48E–02	2.30E–01	1.43E–03	0.00E+00	3.49E–03	6.52E-16	2.86
Alkynes																					
Acetylene	4.2	a	0.42	b	0		7.81E–06	6.36E–06	0.00E+00		7.16E–06	1.88E–02	1.12E–09	0.00E+00	5.90E–02	1.10E–01	1.55E–03	0.00E+00	3.76E–03	7.02E-16	2.19
Aromatics		a																			
Benzene	4.2	a	0.35	b	0.19	d	7.81E–06	5.30E–06	3.72E–04	5.13E–06	7.16E–06	1.88E–02	1.12E–09	1.17E–05	5.90E–02	9.13E–02	1.55E–03	0.00E+00	3.76E–03	7.02E-16	43.74
Toluene	4.2	a	0.48	b	0.19	d	7.81E–06	7.27E–06	3.72E–04		7.16E–06	1.88E–02	1.12E–09	1.17E–05	5.90E–02	1.25E–01	1.55E–03	0.00E+00	3.76E–03	7.02E-16	43.41
O-xylene	4.2	a	0.75	b	0.19	d	7.81E–06	1.14E–05	3.72E–04		7.16E–06	1.88E–02	1.12E–09	1.17E–05	5.90E–02	1.96E–01	1.55E–03	0.00E+00	3.76E–03	7.02E-16	43.87
M-xylene	4.2	a	1.08	b	0.19	d	7.81E–06	1.64E–05	3.72E–04		7.16E–06	1.88E–02	1.12E–09	1.17E–05	5.90E–02	2.82E–01	1.55E–03	0.00E+00	3.76E–03	7.02E-16	44.43
P-xylene	4.2	a	1.03	b	0.19	d	7.81E–06	1.56E–05	3.72E–04		7.16E–06	1.88E–02	1.12E–09	1.17E–05	5.90E–02	2.69E–01	1.55E–03	0.00E+00	3.76E–03	7.02E-16	44.35
Etylbenzene	4.2	a	0.82	b	0.19	d	7.81E–06	1.24E–05	3.72E–04		7.16E–06	1.88E–02	1.12E–09	1.17E–05	5.90E–02	2.14E–01	1.55E–03	0.00E+00	3.76E–03	7.02E-16	43.99
1,2,3-Trime-tylbenzene	4.2	a	0.62	b	0.19	d	7.81E–06	9.39E–06	3.72E–04		7.16E–06	1.88E–02	1.12E–09	1.17E–05	5.90E–02	1.62E–01	1.55E–03	0.00E+00	3.76E–03	7.02E-16	43.65

1,2,4-Trime-tylbenzene	4.2	a	0.81	b	0.19	d	7.81E–06	1.23E–05	3.72E–04		7.16E–06	1.88E–02	1.12E–09	1.17E–05	5.90E–02	2.11E–01	1.55E–03	0.00E+00	3.76E–03	7.02E-16	43.97
1,3,5-Trime-tylbenzene	4.2	a	0.63	b	0.19	d	7.81E–06	9.55E–06	3.72E–04		7.16E–06	1.88E–02	1.12E–09	1.17E–05	5.90E–02	1.64E–01	1.55E–03	0.00E+00	3.76E–03	7.02E-16	43.67
M-ethyl-toluene	4.2	a	1.01	b	0.19	d	7.81E–06	1.53E–05	3.72E–04		7.16E–06	1.88E–02	1.12E–09	1.17E–05	5.90E–02	2.63E–01	1.55E–03	0.00E+00	3.76E–03	7.02E-16	44.31
O-ethyltoluene	4.2	a	0.62	b	0.19	d	7.81E–06	9.39E–06	3.72E–04		7.16E–06	1.88E–02	1.12E–09	1.17E–05	5.90E–02	1.62E–01	1.55E–03	0.00E+00	3.76E–03	7.02E-16	43.65
P-ethyltoluene	4.2	a	0.95	b	0.19	d	7.81E–06	1.44E–05	3.72E–04		7.16E–06	1.88E–02	1.12E–09	1.17E–05	5.90E–02	2.48E–01	1.55E–03	0.00E+00	3.76E–03	7.02E-16	44.21
N-propy-lbenzene	4.2	a	0.79	b	0.19	d	7.81E–06	1.20E–05	3.72E–04		7.16E–06	1.88E–02	1.12E–09	1.17E–05	5.90E–02	2.06E–01	1.55E–03	0.00E+00	3.76E–03	7.02E-16	43.94
I-propy-lbenzene	4.2	a	0.63	b	0.19	d	7.81E–06	9.55E–06	3.72E–04		7.16E–06	1.88E–02	1.12E–09	1.17E–05	5.90E–02	1.64E–01	1.55E–03	0.00E+00	3.76E–03	7.02E-16	43.67
3,5-Dimethyle-thylbenzene	4.2	a	1.32	b	0.19	d	7.81E–06	2.00E–05	3.72E–04		7.16E–06	1.88E–02	1.12E–09	1.17E–05	5.90E–02	3.44E–01	1.55E–03	0.00E+00	3.76E–03	7.02E-16	44.84
3,5-Diethyl-toluene	4.2	a	1.295	c	0.19	d	7.81E–06	1.96E–05	3.72E–04		7.16E–06	1.88E–02	1.12E–09	1.17E–05	5.90E–02	3.38E–01	1.55E–03	0.00E+00	3.76E–03	7.02E-16	44.80
Alcohols																					
Methanol	1.7	a	0.17	b	0.04	d	3.17E–06	2.58E–06	7.84E–05		2.91E–06	7.64E–03	4.54E–10	2.46E–06	2.40E–02	4.43E–02	6.28E–04	0.00E+00	1.53E–03	2.85E-16	9.55
Ethanol	2.3	a	0.53	b	0.04	d	4.41E–06	8.03E–06	7.84E–05		4.05E–06	1.06E–02	6.32E-10	2.46E–06	3.33E–02	1.38E–01	8.73E–04	0.00E+00	2.13E–03	3.97E-16	10.39
Propanol	2.7	a	0.543	c	0.04	d	5.08E–06	8.23E–06	7.84E–05		4.65E–06	1.22E–02	7.26E-10	2.46E–06	3.83E–02	1.42E–01	1.00E–03	0.00E+00	2.45E–03	4.56E-16	10.54
I-propanol	2.7	a	0.4	b	0.04	d	5.08E–06	6.06E–06	7.84E–05		4.65E–06	1.22E–02	7.26E-10	2.46E–06	3.83E–02	1.04E–01	1.00E–03	0.00E+00	2.45E–03	4.56E-16	10.30
Butanol	2.9	a	0.79	b	0.04	d	5.49E–06	1.20E–05	7.84E–05		5.03E–06	1.32E–02	7.85E-10	2.46E–06	4.14E–02	2.06E–01	1.09E–03	0.00E+00	2.64E–03	4.93E-16	11.04
I-butanol	2.9	a	0.375	c	0.04	d	5.49E–06	5.68E–06	7.84E–05		5.03E–06	1.32E–02	7.85E-10	2.46E–06	4.14E–02	9.78E–02	1.09E–03	0.00E+00	2.64E–03	4.93E-16	10.33
s-butanol	2.9	a	0.4	c	0.04	d	5.49E–06	6.06E–06	7.84E–05		5.03E–06	1.32E–02	7.85E-10	2.46E–06	4.14E–02	1.04E–01	1.09E–03	0.00E+00	2.64E–03	4.93E-16	10.37
t-butanol	2.9	a	0.123	c	0.04	d	5.49E–06	1.86E–06	7.84E–05		5.03E–06	1.32E–02	7.85E-10	2.46E–06	4.14E–02	3.21E–02	1.09E–03	0.00E+00	2.64E–03	4.93E-16	9.90
3-pentancl	3.1	a	0.422	c	0.04	d	5.77E–06	6.39E–06	7.84E–05		5.29E–06	1.39E–02	8.25E-10	2.46E–06	4.36E–02	1.10E–01	1.14E–03	0.00E+00	2.78E–03	5.19E-16	10.46
2-Methylbutan-1-ol	3.1	a	0.407	c	0.04	d	5.77E–06	6.17E–06	7.84E–05		5.29E–06	1.39E–02	8.25E-10	2.46E–06	4.36E–02	1.06E–01	1.14E–03	0.00E+00	2.78E–03	5.19E-16	10.44

Table 9.2 contd. …

... Table 9.2 contd.

VOC compound	GWP 100	Ref	POCP	Ref	Particle formation potential	Ref	YLL via climate change	YLL via oxidants	YLL via secondary particles	YLL via cancer	Under-nutrition via climate change	Working capacity via climate change	Diarrhea via climate change	YLD via secondary particles	Crop loss via climate change	Crop loss via oxidants	Meat prod. capacity via climate change	Fish prod. capacity via climate change	Drinking water via climate change	Biodiversity via climate change	Impact value ($/kg)
3-Methylbutan-2-ol	3.1	a	0.366	c	0.04	d	5.77E–06	5.55E–06	7.84E–05		5.29E–06	1.39E–02	8.25E-10	2.46E–06	4.36E–02	9.55E–02	1.14E–03	0.00E+00	2.78E–03	5.19E-16	10.37
2-Methylbutan-2-ol	3.1	a	0.142	c	0.04	d	5.77E–06	2.15E–06	7.84E–05		5.29E–06	1.39E–02	8.25E-10	2.46E–06	4.36E–02	3.70E–02	1.14E–03	0.00E+00	2.78E–03	5.19E-16	9.99
Diacetone alcohol	3.1	a	0.262	c	0.04	d	5.77E–06	3.97E–06	7.84E–05		5.29E–06	1.39E–02	8.25E-10	2.46E–06	4.36E–02	6.83E–02	1.14E–03	0.00E+00	2.78E–03	5.19E-16	10.19
Glycols																					
Ethylene glycol	1.7	a	0.382	c	0.04	d	3.27E–06	5.79E–06	7.84E–05		3.00E–06	7.89E–03	4.69E-10	2.46E–06	2.47E–02	9.97E–02	6.48E–04	0.00E+00	1.58E–03	2.94E-16	9.92
Propylene glycol	2.1	a	0.457	c	0.04	d	4.01E–06	6.92E–06	7.84E–05		3.67E–06	9.65E–03	5.73E-10	2.46E–06	3.03E–02	1.19E–01	7.93E–04	0.00E+00	1.93E–03	3.60E-16	10.19
Ketons																					
Acetone	3.0	a	0.44	b	0		5.64E–06	6.67E–06	0.00E+00		5.17E–06	1.36E–02	8.07E-10	0.00E+00	4.26E–02	1.15E–01	1.12E–03	0.00E+00	2.72E–03	5.07E-16	1.81
Methyl ethyl ketone	3.4	a	0.59	b	0		6.43E–06	8.94E–06	0.00E+00		5.90E–06	1.55E–02	9.20E-10	0.00E+00	4.86E–02	1.54E–01	1.27E–03	0.00E+00	3.10E–03	5.78E-16	2.22
Methyl i-butyl ketone	3.4	a	0.77	b	0		6.43E–06	1.17E–05	0.00E+00		5.90E–06	1.55E–02	9.20E-10	0.00E+00	4.86E–02	2.01E–01	1.27E–03	0.00E+00	3.10E–03	5.78E-16	2.52
Methylpropyl ketone	3.1	a	0.548	c	0		5.90E–06	8.30E–06	0.00E+00		5.41E–06	1.42E–02	8.45E-10	0.00E+00	4.46E–02	1.43E–01	1.17E–03	0.00E+00	2.84E–03	5.31E-16	2.04
Diethylketone	3.1	a	0.414	c	0		5.90E–06	6.27E–06	0.00E+00		5.41E–06	1.42E–02	8.45E-10	0.00E+00	4.46E–02	1.08E–01	1.17E–03	0.00E+00	2.84E–03	5.31E-16	1.82
Methyl-i-propylketone	3.1	a	0.364	c	0		5.90E–06	5.52E–06	0.00E+00		5.41E–06	1.42E–02	8.45E-10	0.00E+00	4.46E–02	9.50E–02	1.17E–03	0.00E+00	2.84E–03	5.31E-16	1.73
Hexan-2-one	3.3	a	0.572	c	0		6.22E–06	8.67E–06	0.00E+00		5.70E–06	1.50E–02	8.89E-10	0.00E+00	4.69E–02	1.49E–01	1.23E–03	0.00E+00	3.00E–03	5.59E-16	2.14
Hexan-3-one	3.3	a	0.599	c	0		6.22E–06	9.08E–06	0.00E+00		5.70E–06	1.50E–02	8.89E-10	0.00E+00	4.69E–02	1.56E–01	1.23E–03	0.00E+00	3.00E–03	5.59E-16	2.19
Methyl-t-butylketone	3.3	a	0.323	c	0		6.22E–06	4.89E–06	0.00E+00		5.70E–06	1.50E–02	8.89E-10	0.00E+00	4.69E–02	8.43E–02	1.23E–03	0.00E+00	3.00E–03	5.59E-16	1.72

Aldehydes																					
Formaldehyde	1.8	a	0.36	b	0		3.38E–06	5.45E–06	0.00E+00		3.10E–06	8.15E–03	4.84E-10	0.00E+00	2.56E–02	9.39E–02	6.70E–04	0.00E+00	1.63E–03	3.04E-16	1.25
Acetaldehyde	2.5	a	0.74	b	0		4.61E–06	1.12E–05	0.00E+00		4.23E–06	1.11E–02	6.60E-10	0.00E+00	3.49E–02	1.93E–01	9.13E–04	0.00E+00	2.22E–03	4.15E-16	2.13
Propionaldehyde	2.8	a	0.86	b	0		5.25E–06	1.30E–05	0.00E+00		4.81E–06	1.27E–02	7.51E-10	0.00E+00	3.97E–02	2.24E–01	1.04E–03	0.00E+00	2.53E–03	4.72E-16	2.45
Butyraldehyde	3.0	a	0.9	b	0		5.64E–06	1.36E–05	0.00E+00		5.17E–06	1.36E–02	8.07E-10	0.00E+00	4.26E–02	2.35E–01	1.12E–03	0.00E+00	2.72E–03	5.07E-16	2.59
I-butyraldehyde	3.0	a	0.59	b	0		5.64E–06	8.94E–06	0.00E+00		5.17E–06	1.36E–02	8.07E-10	0.00E+00	4.26E–02	1.54E–01	1.12E–03	0.00E+00	2.72E–03	5.07E-16	2.07
Valeraldehyde	3.1	a	0.9	b	0		5.90E–06	1.36E–05	0.00E+00		5.41E–06	1.42E–02	8.45E-10	0.00E+00	4.46E–02	2.35E–01	1.17E–03	0.00E+00	2.84E–03	5.31E-16	2.64
Glyoxal	1.9	a	0.19	b	0		3.50E–06	2.88E–06	0.00E+00		3.21E–06	8.43E–03	5.01E-10	0.00E+00	2.64E–02	4.96E–02	6.93E–04	0.00E+00	1.69E–03	3.15E-16	0.98
Methyl-Glyoxal	2.3	a	0.71	b	0		4.23E–06	1.08E–05	0.00E+00		3.88E–06	1.02E–02	6.05E-10	0.00E+00	3.20E–02	1.85E–01	8.37E–04	0.00E+00	2.04E–03	3.80E-16	2.00
Acrolein	2.9	a	0.99	b	0		5.44E–06	1.50E–05	0.00E+00		4.99E–06	1.31E–02	7.78E-10	0.00E+00	4.11E–02	2.58E–01	1.08E–03	0.00E+00	2.62E–03	4.89E-16	2.71
Methanethiol	1.1	a	1.02	b	0		2.12E–06	1.55E–05	0.00E+00		1.94E–06	5.10E–03	3.03E-10	0.00E+00	1.60E–02	2.66E–01	4.19E–04	0.00E+00	1.02E–03	1.90E-16	2.13
Benzaldehyde	3.6	a	0	b	0.19	d	6.70E–06	0.00E+00	3.72E–04		6.15E–06	1.62E–02	9.59E-10	1.17E–05	5.06E–02	0.00E+00	1.33E–03	0.00E+00	3.23E–03	6.03E-16	42.39
Esters																					
Methyl formate	1.8	a	0.033	c	0.04	d	3.38E–06	5.00E–07	7.84E–05		3.10E–06	8.15E–03	4.84E-10	2.46E–06	2.56E–02	8.61E–03	6.70E–04	0.00E+00	1.63E–03	3.04E-16	9.35
Methyl-Acetate	2.2	a	0.068	b	0.04	d	4.12E–06	1.03E–06	7.84E–05		3.77E–06	9.92E–03	5.89E-10	2.46E–06	3.11E–02	1.77E–02	8.14E–04	0.00E+00	1.98E–03	3.70E-16	9.55
ethylacetate	2.5	a	0.24	b	0.04	d	4.61E–06	3.64E–06	7.84E–05		4.23E–06	1.11E–02	6.60E-10	2.46E–06	3.49E–02	6.26E–02	9.13E–04	0.00E+00	2.22E–03	4.15E-16	9.94
n-Propyl acetate	2.6	a	0.29	c	0.04	d	4.98E–06	4.39E–06	7.84E–05		4.56E–06	1.20E–02	7.12E-10	2.46E–06	3.76E–02	7.57E–02	9.85E–04	0.00E+00	2.40E–03	4.47E-16	10.09
i-Propyl acetate	2.6	a	0.21	c	0.04	d	4.98E–06	3.18E–06	7.84E–05		4.56E–06	1.20E–02	7.12E-10	2.46E–06	3.76E–02	5.48E–02	9.85E–04	0.00E+00	2.40E–03	4.47E-16	9.95
n-butylacetate	2.8	a	0.44	b	0.04	d	5.25E–06	6.67E–06	7.84E–05		4.81E–06	1.27E–02	7.51E-10	2.46E–06	3.97E–02	1.15E–01	1.04E–03	0.00E+00	2.53E–03	4.72E-16	10.40
2-Ethylhexyl acetate	3.1	a	0.3	b	0.04	d	5.90E–06	4.55E–06	7.84E–05		5.41E–06	1.42E–02	8.45E-10	2.46E–06	4.46E–02	7.83E–02	1.17E–03	0.00E+00	2.84E–03	5.31E-16	10.28
Ethers																					
Dimethylether	2.3	a	0.33	b	0		4.41E–06	5.00E–06	0.00E+00		4.05E–06	1.06E–02	6.32E-10	0.00E+00	3.33E–02	8.61E–02	8.73E–04	0.00E+00	2.13E–03	3.97E-16	1.39
Diethyl-Ether	2.9	a	0.52	b	0		5.49E–06	7.88E–06	0.00E+00		5.03E–06	1.32E–02	7.85E-10	0.00E+00	4.14E–02	1.36E–01	1.09E–03	0.00E+00	2.64E–03	4.93E-16	1.92

Table 9.2 contd. …

... *Table 9.2 contd.*

VOC compound	GWP 100	Ref	POCP	Ref	Particle formation potential	Ref	YLL via climate change	YLL via oxidants	YLL via secondary particles	YLL via cancer	Under-nutrition via climate change	Working capacity via climate change	Diarrhea via climate change	YLD via secondary particles	Crop loss via climate change	Crop loss via oxidants	Meat prod. capacity via climate change	Fish prod. capacity via climate change	Drinking water via climate change	Biodiversity via climate change	Impact value ($/kg)
Methyl-t-Butyl-Ether	3.1	a	0.49	b	0		5.77E–06	7.42E–06	0.00E+00		5.29E–06	1.39E–02	8.25E-10	0.00E+00	4.36E–02	1.28E–01	1.14E–03	0.00E+00	2.78E–03	5.19E-16	1.92
Diisopropylether	3.2	a	0.476	c	0		5.97E–06	7.21E–06	0.00E+00		5.48E–06	1.44E–02	8.54E-10	0.00E+00	4.51E–02	1.24E–01	1.18E–03	0.00E+00	2.88E–03	5.37E-16	1.94
Ethyl-t-butylether	3.2	a	0.214	c	0		5.97E–06	3.24E–06	0.00E+00		5.48E–06	1.44E–02	8.54E-10	0.00E+00	4.51E–02	5.58E–02	1.18E–03	0.00E+00	2.88E–03	5.37E-16	1.49
Alcohol and glycol ethers		a																			
2-Methoxye-thanol	2.1	a	0.3	c	0.04	d	4.01E–06	4.55E–06	7.84E–05		3.67E–06	9.65E–03	5.73E-10	2.46E–06	3.03E–02	7.83E–02	7.93E–04	0.00E+00	1.93E–03	3.60E-16	9.92
2-Ethoxye-thanol	2.4	a	0.387	c	0.04	d	4.51E–06	5.86E–06	7.84E–05		4.14E–06	1.09E–02	6.46E-10	2.46E–06	3.41E–02	1.01E–01	8.93E–04	0.00E+00	2.17E–03	4.06E-16	10.17
1-Butoxypropanol	2.9	a	0.436	c	0.04	d	5.38E–06	6.61E–06	7.84E–05		4.94E–06	1.30E–02	7.70E-10	2.46E–06	4.07E–02	1.14E–01	1.07E–03	0.00E+00	2.59E–03	4.84E-16	10.41
2-Butoxye-thanol	2.7	a	0.438	c	0.04	d	5.16E–06	6.64E–06	7.84E–05		4.73E–06	1.24E–02	7.39E-10	2.46E–06	3.90E–02	1.14E–01	1.02E–03	0.00E+00	2.49E–03	4.64E-16	10.38
1-Methoxy-2-propanol	2.4	a	0.368	c	0.04	d	4.51E–06	5.58E–06	7.84E–05		4.14E–06	1.09E–02	6.46E-10	2.46E–06	3.41E–02	9.60E–02	8.93E–04	0.00E+00	2.17E–03	4.06E-16	10.13
Carboxylic acids																					
Formic acid	1.2	a	0.032	c	0		2.21E–06	4.85E–07	0.00E+00		2.02E–06	5.32E–03	3.16E-10	0.00E+00	1.67E–02	8.35E–03	4.37E–04	0.00E+00	1.06E–03	1.98E-16	0.47
Acetic Acid	1.8	a	0.15	b	0		3.38E–06	2.27E–06	0.00E+00		3.10E–06	8.15E–03	4.84E-10	0.00E+00	2.56E–02	3.91E–02	6.70E–04	0.00E+00	1.63E–03	3.04E-16	0.89
Propanoic acid	2.6	a	0.15	c	0		4.91E–06	2.27E–06	0.00E+00		4.50E–06	1.18E–02	7.03E-10	0.00E+00	3.71E–02	3.91E–02	9.72E–04	0.00E+00	2.37E–03	4.42E-16	1.18
Halocarbons																					
Methyl chloride	12.0	a	0.005	c	0		2.26E–05	7.58E–08	0.00E+00		2.07E–05	5.44E–02	3.23E–09	0.00E+00	1.70E–01	1.30E–03	4.46E–03	0.00E+00	1.09E–02	2.03E-15	4.27
Methylene chloride	9.0	a	0.068	c	0		1.69E–05	1.03E–06	0.00E+00		1.55E–05	4.08E–02	2.42E–09	0.00E+00	1.28E–01	1.77E–02	3.35E–03	0.00E+00	8.15E–03	1.52E-15	3.31
Chloroform	16.0	a	0.023	c	0		3.01E–05	3.49E–07	0.00E+00		2.76E–05	7.25E–02	4.30E–09	0.00E+00	2.27E–01	6.00E–03	5.95E–03	0.00E+00	1.45E–02	2.70E-15	5.72

1,1-dichloroe-thane	1.0	a	0.079	b	0		1.88E–06	1.20E–06	0.00E+00		1.72E–06	4.53E–03	2.69E-10	0.00E+00	1.42E–02	2.06E–02	3.72E–04	0.00E+00	9.06E–04	1.69E-16	0.49
1,1-dichloroe-thylene	1.0	a	0.025	b	0		1.88E–06	3.79E–07	0.00E+00		1.72E–06	4.53E–03	2.69E-10	0.00E+00	1.42E–02	6.52E–03	3.72E–04	0.00E+00	9.06E–04	1.69E-16	0.40
1,2-Dichloroe-thylene	1.0	a	0.42	b	0		1.88E–06	6.36E–06	0.00E+00		1.72E–06	4.53E–03	2.69E-10	0.00E+00	1.42E–02	1.10E–01	3.72E–04	0.00E+00	9.06E–04	1.69E-16	1.07
Tetrachloroe-thylene	4.5	a	0.014	b	0		8.46E–06	2.12E–07	0.00E+00		7.76E–06	2.04E–02	1.21E–09	0.00E+00	6.39E–02	3.65E–03	1.67E–03	0.00E+00	4.08E–03	7.61E-16	1.62
Trichloroe-thylene	4.5	a	0.09	b	0		8.46E–06	1.36E–06	0.00E+00		7.76E–06	2.04E–02	1.21E–09	0.00E+00	6.39E–02	2.35E–02	1.67E–03	0.00E+00	4.08E–03	7.61E-16	1.75
Methyl chloroform	160.0	a	0.003	b	0		3.01E–04	4.55E–08	0.00E+00		2.76E–04	7.25E–01	4.30E–08	0.00E+00	2.27E+00	7.83E–04	5.95E–02	0.00E+00	1.45E–01	2.70E-14	56.81
Others																					
Methyl Mercaptan	4.5	a	0.62	b	0		8.46E–06	9.39E–06	0.00E+00		7.76E–06	2.04E–02	1.21E–09	0.00E+00	6.39E–02	1.62E–01	1.67E–03	0.00E+00	4.08E–03	7.61E-16	2.65
Diethyl Sulfide	4.5	a	0.62	b	0		8.46E–06	9.39E–06	0.00E+00		7.76E–06	2.04E–02	1.21E–09	0.00E+00	6.39E–02	1.62E–01	1.67E–03	0.00E+00	4.08E–03	7.61E-16	2.65
Diethyl Disulfide	4.5	a	0.36	b	0		8.46E–06	5.45E–06	0.00E+00		7.76E–06	2.04E–02	1.21E–09	0.00E+00	6.39E–02	9.39E–02	1.67E–03	0.00E+00	4.08E–03	7.61E-16	2.21

The uncertainty in the estimation of the environmental impact indicator is caused by varying local conditions (e.g., population density), potency of SOA relative PM2.5 and varying particle formation efficiency for individual chemical compounds. A factor of 2 is assumed.

9.3.4 YLL via cancer

There are four VOCs that are known carcinogens: ethene, propene, benzene and 1,3-butadiene.

For **benzene**, the risk for leukemia is estimated to be 5.9 cases per million among people who experience lifelong exposure to benzene concentrations of 1 μg/m^3 in the air [4]. The mortality in leukemia is about 50% [5]. A study of personal exposure to benzene in air in six European cities showed an average exposure to about 15 μg/m^3, which was shown to be twice the ambient air exposure [6]. Emissions to ambient air of benzene thus causes 0.50* 5.9/1000000*7200000000*15/2*1/75 = 2.12E+3 fatal cases per year and 2.87E+04YLLs per year with an expected reduction in life expectancy of 13.5 years per case. The value 13.5 YLL is based on data from UK (13.6 YLL) [7], from Norway (13.5 YLL) [8] and from Iran (13.5) [9].

The global emission of benzene is estimated to be 5.6 TgC/year by Henze et al. [2]. The average YLLs/kg benzene is therefore 2.87E+4/5.6E+09 = 5.13E–06 YLL/kg benzene.

For **ethene,** Victorin uses USEPA provisional estimates of the lifetime cancer risk for ethylene-oxide of 1E–04 per μg/m^3 and assumes a metabolization rate of 5% to estimate the cancer risk for ethylene [10]. This will give a risk estimate for ethene of 5E–06 per μg/m^3, which is 0.85 times the risk of benzene if assuming the same dispersion patterns as benzene. While benzene mainly causes leukemia, ethene is associated with lung cancer. Mortality in lung cancer is about 80% [11]. The average YLL per case of fatal lung cancer is 11.8 YLL in UK [7], 14 YLL in Norway [8] and (9.5) Iran [9]. A best estimate for the global average of 10 YLL is assumed. The environmental impact factor then becomes 0.8*5.13E–06*0.85*10/13.5 = 2.58E–06 YLL/kg ethene.

For **propylene,** Victorin [10] uses USEPA estimates of the lifetime cancer risk for propylene-oxide for which the lifetime inhalation unit risk is 3.7E–06 per μg/m^3 and assumes a metabolization rate of 10% to estimate the cancer risk for propylene. This will give an estimate that is 0.063 times the risk of benzene if assuming the same dispersion patterns as benzene. While benzene mainly causes leukemia, propylene is associated with lung cancer. Mortality in lung cancer is about 80% [11]. The average YLL per case of fatal lung cancer is 11.8 YLL in UK [7], 14 YLL in Norway

[8] and (9.5) Iran [9]. A best estimate for the global average of 10 YLL is assumed. The environmental impact factor then becomes 0.8*5.13E–06 *0.063*10/13.5 = 1.92E–07 YLL/kg propylene.

USEPA estimates the inhalation unit risk for **1,3-butadiene** exposure to 3×10^{-5} per μg/m^3. Assuming similar dispersion and exposure patterns as for benzene, the risk for cancer caused by an emission of 1 kg 1,3-butadiene is 30/5.9 = 5.08 times the benzene risk. While benzene mainly causes leukemia, propylene is assumed to be associated with lung cancer. Mortality in lung cancer is about 80% [11]. The average YLL per case of fatal lung cancer is 11.8 YLL in UK [7], 14 YLL in Norway [8] and (9.5) Iran [9]. A best estimate for the global average of 10 YLL is assumed. The environmental impact factor then becomes 0.8*5.13E–06 *5.08*10/13.5 = 1.54E–05 YLL/kg 1,3-butadiene.

The uncertainty of the environmental impact factors is mainly caused by variations in exposure conditions. The extent to which emissions is inhaled varies considerably. An uncertainty factor of 3 is assumed.

9.3.5 Undernutrition via climate change

The environmental impact factor for undernutrition from CO_2 was 1.72E–06 personyears/kg CO_2 (6.1.3.5). If GWP for a VOC is denoted GWP_i, the environmental impact factor for the ith compound is 1.72E–06*GWP_i. Environmental impact factors for different VOC calculated is this way are shown in Table 9.2.

The uncertainty in the environmental impact factor for undernutrition from CO_2 was estimated to be 2 in Sec. 6.1.3.5. An additional uncertainty due to varying local and regional conditions of a factor of 2 is assumed, resulting in a total uncertainty of exp $\sqrt{(\ln 2)^2 + (\ln 2)^2} = 2.7$.

9.3.6 Working capacity via climate change

The environmental impact factor for loss of working capacity from CO_2 via climate change was 4.53E–03 personhours/kg CO_2 (6.1.3.6). If GWP for a VOC is denoted GWP_i, the environmental impact factor for the ith compound is 4.53E–03*GWP_i. Environmental impact factors for different VOC calculated in this way are shown in Table 9.2.

The uncertainty in the environmental impact factor for working capacity from CO_2 was estimated to be 2 in Sec. 6.1.3.6. An additional uncertainty due to varying local and regional conditions of a factor of 2 is assumed, resulting in a total uncertainty of exp $\sqrt{(\ln 2)^2 + (\ln 2)^2} = 2.7$.

9.3.7 Diarrhea via climate change

The environmental impact factor for diarrhea from CO_2 via climate change is 2.69E–10 personyears/kg CO_2 (6.1.3.7). If GWP for a VOC is denoted GWP_i, the environmental impact factor for the ith compound is 2.69E–10*GWP_i. Environmental impact factors for different VOC calculated in this way are shown in Table 9.2.

The uncertainty in the environmental impact factor for diarrhea from CO_2 via climate change was estimated to be 3 in Sec. 6.1.3.7. An additional uncertainty due to varying local and regional conditions of a factor of 2 is assumed, resulting in a total uncertainty of exp $\sqrt{(\ln 3)^2 + (\ln 2)^2} = 3.7$.

9.3.8 YLD via secondary particles

The environmental impact factor was determined in Sec. 7.1.3.1 to be 6.16E–05 YLD/kg PM2.5.

The environmental impact factor for VOC's impact on YLD via secondary particles is 6.16E–05*P YLL/kg VOC measured as carbon, where P is the particle formation efficiency. Environmental impact factors for different VOCs are shown in Table 9.2.

The uncertainty in the estimation of the environmental impact indicator is caused by varying local conditions (e.g., population density), potency of SOA relative PM2.5 and varying particle formation efficiency for individual chemical compounds. The same uncertainty, a factor of 3.2, is assumed as for PM2.5's impact on YLD (7.1.3.3).

9.3.9 Crop loss via climate change

The aggregated environmental impact factor for crop loss from CO_2 and via climate change is 1.42E–02 kg crop/kg CO_2 (6.1.3.8 and 6.1.3.9). If GWP for a VOC is denoted GWP_i, the environmental impact factor for the ith compound is 1.42E–02*GWP_i. Environmental impact factors for different VOC calculated in this way are shown in Table 9.2.

The uncertainty in the environmental impact factor for crop loss from CO_2 via climate change was estimated to 3 in Secs. 6.1.3.8 and 6.1.3.9. An additional uncertainty due to varying local and regional conditions of a factor of 2 is assumed, resulting in a total uncertainty of exp $\sqrt{(\ln 3)^2 + (\ln 2)^2} = 3.7$.

9.3.10 Crop loss via oxidants

The environmental impact factor for crop loss from NO_x via tropospheric ozone was determined to be 0.120 kg crop/kg NO_x in 6.3.3.8. The POCP used for NO_x was 0.46. If the POCP for the ith compound is denoted $POCP_i$, then the environmental impact factor for the ith VOC is 0.120/0.46*$POCP_i$. Environmental impact factors for different VOC calculated in this way are shown in Table 9.2.

The uncertainty in the environmental impact factor for crop loss from NO_x via oxidants was estimated as 2 in Sec. 6.3.3.8. The same uncertainty is assumed for VOC, with an additional uncertainty due to varying local and regional conditions of a factor of 2, resulting in a total uncertainty of $\exp\sqrt{(\ln 2)^2 + (\ln 2)^2} = 2.7$.

9.3.11 Production capacity of meat via climate change

The environmental impact factor for decreased production capacity of meat from CO_2 and via climate change is 3.72E–04 kg meat/kg CO_2 (6.1.3.10). If GWP for the ith VOC is denoted GWP_i, the environmental impact factor for the ith compound is 3.72E–04*GWP_i. Environmental impact factors for different VOC calculated in this way are shown in Table 9.2.

The uncertainty in the environmental impact factor for decreased meat production capacity from CO_2 via climate change was estimated as 3 in Sec. 6.1.3.10. An additional uncertainty due to varying local and regional conditions of a factor of 2 is assumed, resulting in a total uncertainty of $\exp\sqrt{(\ln 3)^2 + (\ln 2)^2} = 3.7$.

9.3.12 Production capacity of fish via climate change

The environmental impact factor for decreased production capacity of fish from CO_2 and via climate change is caused by ocean acidification. As mentioned in Sec. 9.1, VOCs corresponding to 400 Tg C out of the total 1350 Tg C is oxidized to CO_2. It is assumed that this is chiefly methane. The total global emission of methane is around 600 Tg C. The environmental impact factor for CO_2 on fish production capacity was determined in Sec. 6.1.3.11 to 2.92E–05 kg fish/kg CO_2. The environmental impact factor for methane on production capacity of fish via climate change then becomes 2.92E–05*400/600*12/16 = 1.46E–05 kg fish/kg methane.

The uncertainty in the environmental impact factor for decreased fish production capacity from CO_2 via climate change was estimated to be 2 in Sec. 6.1.3.11. As methane has a residence time of around ten years, no additional uncertainty due to varying local and regional conditions of a factor is assumed, resulting in a total uncertainty of 2.

9.3.13 Production capacity of drinking water via climate change

The environmental impact factor for decreased production capacity of drinking water from CO_2 and via climate change is 9.06E–04 m^3 drinking water/kg CO_2 (6.1.3.13). If GWP for the ith VOC is denoted GWP_i, the environmental impact factor for the ith compound is 9.06E–04*GWP_i. Environmental impact factors for different VOC calculated in this way are shown in Table 9.2.

As for CO_2, the uncertainty is high, but as the 9.06E–04 m^3/kg CO_2 is an upper limit, the uncertainty is only towards lower values.

9.3.14 Decreased biodiversity via climate change

The environmental impact factor for biodiversity from CO_2 and via climate change is 1.69E–16 shares of the present impact on biodiversity per kg of CO_2 (Sec. 6.1.3.14). If GWP for the ith VOC is denoted GWP_i, the environmental impact factor for the ith compound is 1.69E–16* GWP_i. Environmental impact factors for different VOC calculated in this way are shown in Table 9.2.

The uncertainty is assumed to be the same as for CO_2, a factor of 4.

6.4 Summary and Monetary Valuation of VOC Emissions

Environmental impact factors and monetary values for individual VOCs are shown in Table 9.2.

It is striking that VOCs with high particle formation potential get very high monetary values. Considering that the modelling does not include differences in exposure pattern between monitored PM2.5 in air and SOA, the need for improved modelling is obvious.

The average environmental impact value for NMVOCs in Table 9.2 is 10.1 \$/kg.

References

1. Hallquist, M. et al., The formation, properties and impact of secondary organic aerosol: current and emerging issues. *Atmos. Chem. Phys.*, 2009. **9**(14): 5155–5236.
2. Henze, D.K. et al., Global modelling of secondary aerosol formation from aromatic hydrocarbons: high- vs. low-yield pathways. *Atmospheric Chemistry and Physics*, 2008. **8**: 2405–2421.
3. Labouze, E. et al., Photochemical ozone creation potentials. *The International Journal of Life Cycle Assessment*, 2004. **9**(3): 187–195.
4. WHO, Exposure to benzene: a major public health concern, in: *Preventing Disease Through Healthy Environments*. 2010, WHO: Geneva.

5. Veisani, Y., S. Khazaei and A. Delpisheh, 5-year survival rate based on the types of leukemia in Iran: a Meta-analysis. *Caspian Journal of Internal Medicine*, 2018. **9**(4): 316–324.
6. Cocheo, V. et al., Urban benzene and population exposure. *Nature*, 2000. **404**(6774): 141–142.
7. Burnet, N.G. et al., Years of life lost (YLL) from cancer is an important measure of population burden—and should be considered when allocating research funds. *British Journal of Cancer*, 2005. **92**(2): 241–245.
8. Brustugun, O.T., Møller, B. and Å. Helland, Years of life lost as a measure of cancer burden on a national level. *Br. J. Cancer*, 2014. **111**(5): 1014–1020.
9. Mirzaei, M., M. Mirzadeh and M. Mirzaei, Expected Years of Life Lost Due to Adult Cancer Mortality in Yazd (2004–2010). *Asian Pac. J. Cancer Prev.*, Cancer Control in Western Asia Special Issue, 2016. **17**: 101–105.
10. Victorin, K., *Risk assessment of carcinogenic air pollutants*. 1998, Karolinska Institutet: Stockholm.
11. Matsuda, A. and K. Katanoda, Five-Year Relative Survival Rate of Lung Cancer in the USA, Europe and Japan. *Japanese Journal of Clinical Oncology*, 2013. **43**(12): 1287–1288.
12. Altenstedt, J. and K. Pleijel, An Alternative Approach to Photochemical Ozone Creation Potentials Applied Under European Conditions. *Journal of the Air and Waste Management Association*, 2000. **50**(6): 1023–1036.
13. Derwent, R.G. et al., Photochemical ozone creation potentials for organic compounds in northwest Europe calculated with a master chemical mechanism. *Atmospheric Environment*, 1998. **32**(14): 2429–2441.

10

Emissions to Water

10.1 BOD to Freshwater

10.1.1 Specification of the emission

BOD means biological oxygen demand. It is a measure of substances in effluents or in recipients that contributes to oxygen deficiency in waters. It reflects the amount of oxygen needed for decomposing substances, mostly organic, in water. It is defined by a standardized measurement method, such as from the USGS [1]. Oxygen demand may be determined during 5 or 7 days. Sometimes you see BOD reported as BOD_5 or BOD_7. There are also variants that measures only decomposition of carbonaceous material (CBOD) or a combination of carbonaceous and nitrogenous materials. Often it is unclear which BOD-method that was used, which means that there is an uncertainty of about 30% about which values that are reported. Another measure of the oxygen demand is COD, chemical oxygen demand. COD is typically twice as high as BOD. COD is not modelled here. It would mean double counting to value both BOD and COD. Besides, a model of COD's impacts would require determining how much BOD it would generate.

The emission valued is any BOD to freshwater anywhere in the world in 2015.

10.1.2 Impacts pathways

BOD emissions to freshwater cause oxygen deficiency, which decreases fish production capacity and biodiversity. The extent of the impacts varies locally depending on water flows, temperature and amount emitted. A part of the BOD emitted to freshwater is transferred via rivers to seawater, but its impact on seawater organisms is not modelled. It is believed to be too small to motivate the resources needed for modelling.

10.1.3 Environmental impact factors

10.1.3.1 Decreased fish production capacity in freshwater from oxygen deficiency

UN global environmental statistics from GEMSTAT [2] show that Dissolved Oxygen (DO) levels are less than 9.5 mg/l (which cause reduced production capacity in lakes with an average temperature below 20°C), in 62% of the monitoring sites, and less than 6 mg/l (which cause reduced production capacity in lakes with an average temperature above 20°C) in 12% of the monitoring sites. 63% of the lakes have temperatures below 20°C and the rest 37% lies above 20°C. This allows an approximate estimate of the share of lakes with oxygen deficiency to 0.12 + (0.62–0.12)*0.63 = 0.43 or 43% of the lakes. The growth rate of fish is assumed to be directly proportional to DO at these levels. The DO levels below 9.5 is as an average 7.5 mg/l, indicating a decreased fish growth rate of (9.5–7.5)/9.5 = 0.21 for the < 20-degree lakes with oxygen deficiency covering 0.63*0.12 + 0.63*0.50 = 39% of the total lake area. The DO levels below 6 mg/l areas an average 5 mg/l in > 20 degrees lakes, indicating a decreased fish growth of (6–5)/6 or 16.7% in 0.37*0.12 = 4.4% of the total lake area. The total global freshwater area is 5988250 km^2 (http://en.worldstat.info). The average fish growth capacity is assumed to be 1000 kg/km^2, year. This means a total loss of production capacity of (0.21*0.39 + 0.167*0.044)*1000*5988250 kg/year = 5.34E+08 kg/year.

The contribution of BOD to oxygen deficiency depends not only on BOD emissions but also on N and P emissions. P and N stimulate growth of phytoplankton which is at the base of our food chain. The Redfield ratio (C:N:P = 106:16:1) for phytoplankton is often used to model the growth of biomass. When decaying, biomass requires oxygen. In environment where growth is restricted by access to P, like in most freshwaters, 1 molecule of P creates 106 molecules of biomass C. 1 kg of BOD will therefore correspond to 0.00737 kg P in terms of DO depletion. 0.00737 is the ratio of the mass of 1 molecule of P relative to the 127.25 molecules of O_2 that is needed to oxidize 106 molecules of C, 16 molecules of N and one molecule of P. So, the contribution of 1 kg of BOD to eutrophicated freshwater depends on the BOD emissions and P emissions.

Total global BOD emission to freshwater estimated from country specific World Bank data for 2004–2007 is estimated as 1.6E+10 kg. This estimation is based on an assumption that per capita emissions from the countries not covered by World Bank statistics (46% of the population) are the same as those covered. Global P-tot emissions to freshwater is estimated 3.8E+10 kg/yr [3].

Therefore, the contribution from 1 kg BOD to DO depletion in freshwaters are 0.00737/(1.6E+10*0.00737 + 3.8E+10) = 1.93E–13.

The environmental impact factor then becomes 5.34E+08*1.93E–13 = 1.03E–04 kg fish/kg BOD.

The model builds on several assumptions: (1) Extrapolation of DO levels to waters not monitored, (2) independency between temperature and DO levels, (3) a linear relation between DO deficiency and decreased growth rate and (4) average production capacity of 1000 kg/km^2 (which is relevant for temperate regions). Therefore, a standard deviation of a factor of 2 is assumed.

10.1.3.2 Decreased biodiversity due to oxygen deficiency

According to IUCN red-list (http://www.iucnredlist.org/search, accessed at 15 October 2014), 0.007 of red-listed species are threatened by eutrophication. The contribution from BOD is the same as in 10.1.3.1, 1.93E–13 per kg BOD. The environmental impact factor is 0.007*1.93E–13 = 1.35E–15 shares of threat to biodiversity per kg of BOD.

It is unclear how well the data reported to IUCN cover the real situation. An uncertainty of a factor of 3 is assumed.

10.1.4 Summary and monetary valuation of BOD emissions to freshwater

Environmental impact factors and monetary values for emissions of BOD to freshwaters are shown in Table 10.1. A conclusion from the assessment of monetary impact values of BOD emissions to freshwater is that this is a moderate economic problem as an average. An efficient abatement strategy should therefore focus on local conditions and be dealt with for each source.

10.2 N-tot to Freshwater

10.2.1 Specification of the emission

N-tot denotes total bounded nitrogen. Free nitrogen, N_2, is not included. Some nitrogen is organically bounded, and some are inorganic compounds.

The emission valued here is N-tot to freshwater anywhere in the world 2015.

Table 10.1. Environmental impact factors and monetary impact values for BOD emissions to freshwater.

Environmental good	Impact indicator	Unit	Pathway	Environmental impact factor	Uncertainty	Indicator value ($/unit)	Impact value ($/kg BOD)
fish	production capacity	kg	oxygen deficiency	1.03E–04	2	2	2.49E–04
biodiversity	share of threat to redlisted species	dimensionless	oxygen deficiency	1.35E–15	3	7.61E+10	1.03E–04
						sum	3.52E–04

10.2.2 Impacts pathways

Emissions of N-tot to water stimulate growth of various organisms, such as fish, if sufficient oxygen is available. When there is an oxygen deficiency, the organisms die. In areas with oxygen deficiency fish production capacity and biodiversity decrease. In areas with sufficient oxygen, N-tot increases fish production capacity. P is mostly rate limiting in freshwaters, why N-tot emissions only cause oxygen deficiency and increased growth of organisms in sea waters. So, the environmental impact from N-tot emissions is caused by the N-tot that is transferred to the sea via rivers.

10.2.3 Environmental impact factors

10.2.3.1 Decrease of fish production capacity due to oxygen deficiency

According to Diaz and Rosenberg [4], dead zones in sea cover 245 000 km^2. A typical production rate of 10 kg/ha and year will give a total loss of 2.45E+08 kg/year. FAO has recommended that the global catch should level out at 100 million tons for fishing to be sustainable. This corresponds to about 3 kg per ha if all ocean and sea areas are included. Assuming 10 kg/ha for production and fishing at continental shelves may thus be of a reasonable order.

The contribution to the total loss of fish production capacity from 1 kg N-tot depends not only on the total N-tot emissions but also on the total BOD-emission. Galloway et al. [5] estimate the input to rivers in the 1990s of N-tot to 118.1 Tg/yr and the riverine flux to oceans to 47.8 Tg/yr. This gives an average transfer efficiency of 40%. BOD emissions are estimated to 0.83 Tg/yr N-eq. 1 kg of N-tot emitted to freshwater is thus contributing to 1/(118.1E+09 + 0.83E+09) = 8.41E–12 to the decrease in fish productivity.

The environmental impact factor is then 2.45E+08*8.41E–12 = 2.06E–03 kg fish/kg N-tot.

As for the emission of BOD to freshwater, the model builds on several assumptions: (1) Extrapolation of DO levels to waters not monitored, (2) independency between temperature and DO levels, (3) a linear relation between DO deficiency and decreased growth rate and (4) average production capacity of 1000 kg/km^2 (which is relevant for temperate regions). Therefore, a standard deviation of a factor of 2 is estimated.

10.2.3.2 Increase of fish production capacity due to fertilization

As P is rate limiting for growth in freshwaters, the increase in fish production capacity by fertilization is limited to that caused by nitrogen that is transferred to seawater via rivers.

The total global production capacity for fish can be estimated from FAO guidelines. FAO has recommended that the global catch should level out at 100 million tonnes for fishing to be sustainable. The global riverine flux of N-tot to oceans is estimated to be 47.8Tg/year [5]. If mixed evenly in the upper 100 meters, this will contribute to a 0.148% increase of reactive nitrogen. (The ocean area is 360 million km^2, the average concentration of reactive nitrogen is 0.9 g/m^3 and the added concentration from atmospheric deposition is 0.00133 g/m^3 in the upper 100 m.) It is here assumed that the increase in fish catch is proportional to nitrogen availability, as nitrogen is rate limiting. (This is a rough simplification as many other factors determine which catch is available.) The extent of the increased fish production capacity due to anthropogenic nitrogen is therefore 1E+11*0.00133/0.9 = 1.48E+08 kg.

Galloway et al. [5] estimate the N-tot input to rivers in the 1990s to 118.1 Tg/yr. The environmental impact factor is thus 1.48E+08/1.181E+11 = 1.25E–03 kg fish/kg N-tot.

The uncertainty in this model is large. It is a very primitive model ignoring the dynamics and complexity of ecosystem models and the mixing processes in estuaries and coastal regions. An uncertainty of a factor of 10 is assumed.

10.2.3.3 Decreased biodiversity from oxygen deficiency

IUCN red-list http://www.iucnredlist.org/search, accessed at 15 October 2014 estimates that 0.007 of all red-listed species are threatened in aquatic environments from effluents from agriculture and forestry.

The same contribution to decreased biodiversity is assumed as for decreased fish production capacity due to oxygen deficiency, i.e., 8.41E–12 per kg N-tot. The environmental impact factor is then 7E–04*8.41E–12 = 5.89E–15 shares of threat to red-listed species per kg of N-tot.

It is unclear how well the data reported to IUCN cover the real situation. An uncertainty of a factor of 3 is assumed.

10.2.4 Summary and monetary valuation of N-tot emissions to freshwater

Environmental impact factors and monetary values for emissions of N-tot to freshwaters are shown in Table 10.2. Although the uncertainty in valuing the environmental impact of N-tot fertilization is large, the monetary values of N-tot emissions are unlikely to be a significant issue in most product or process development.

Table 10.2. Summary of environmental impact factors and monetary impact values of emission of 1 kg N-tot to freshwater.

Environmental good	Impact indicator	Unit	Pathway	Environmental impact factor	Uncertainty	Indicator value ($/unit)	Impact value ($/kg N-tot)
fish	production capacity	kg	oxygen deficiency	2.06E–03	2	2.42	4.99E–03
fish	production capacity	kg	fertilization	–1.25E–03	10	2.42	–3.03E–03
biodiversity	share of threat to redlisted species	dimensionless	oxygen deficiency	5.89E–15	3	7.61E+10	4.48E–04
						sum	2.40E–03

10.3 N-tot to Seawater

As only 40% of N-tot is emitted to freshwater reaches seawater, emissions directly to sea water will cause impacts that are 2.5 times higher. Its impact value will then be 2.40E–3*1/0.4 = 6.00E–03 $/kg N-tot.

10.4 P-tot to Freshwater

10.4.1 Specification of the emission

There are several P-containing compounds emitted to water, but they are seldom identified individually in emission data reports. Sometimes organic P and inorganic P is separately identified. While N is a relatively water soluble and mobile nutrient P compounds are less soluble and mobile.

P-tot is emitted from agriculture and from waste water treatment plants.

The emission valued here is P-tot to freshwater anywhere is the world 2015.

10.4.2 Impacts pathways

Emissions of P-tot to water stimulates growth of various organisms, such as fish, if sufficient oxygen is available. When there is an oxygen deficiency, the organisms die. In areas with oxygen deficiency fish production capacity and biodiversity decrease. In areas with sufficient oxygen, P-tot increases fish production capacity. P is mostly growth rate limiting in freshwaters, so impacts from P-tot emissions occur primarily in freshwaters, and not in sea waters. P-tot emissions decrease fish production capacity and biodiversity by causing oxygen deficiency.

10.4.3 Environmental impact factors

10.4.3.1 Decreased fish production capacity via oxygen deficiency in fresh waters

UN global environmental statistics from GEMSTAT[2] show that Dissolved Oxygen (DO) levels are less than 9.5 mg/l (which cause reduced production capacity in lakes with an average temperature below 20°C), in 62% of the monitoring sites and less than 6 mg/l (which cause reduced production capacity in lakes with an average temperature above 20°C) in 12% of the monitoring sites. 63% of the lakes have temperatures below 20°C and the rest 37% lies above 20°C. This allows an approximate estimate of the share of lakes with oxygen deficiency to be 0.12*0.37 + 0.63*0.12 + 0.63*0.45 = 0.40 or 40% of the lakes. The growth rate of fish is assumed to be directly proportional to DO at these levels. The DO levels below 9.5 and 6 is as an average 7.5 mg/l indicating a decreased fish growth rate of 2/9.5 = 0.21 for the < 20-degree lakes covering 36% of the area (0.63*0.12 + 0.63*0.45) and below 5 in > 20 degrees lakes indicating a decreased fish growth of 1/6 or 16.7% in 4.4% (0.37*0.12) of the lake areas. The total global freshwater area is 5988250 km^2 (http://en.worldstat.info). The average fish growth capacity is assumed to be 1000 kg/km^2, year. This means a total loss of production capacity of (0.21*0.36 + 0.167*0.044)*1000*5988250 kg/year = 4.97E+08 kg/year.

The contribution of P-tot to oxygen deficiency depends not only on P-tot emissions but also on N and BOD emissions. P and N stimulate growth of phytoplankton which is at the base of the food chain. The Redfield ratio (C:N:P = 106:16:1) for phytoplankton is often used to model the growth of biomass. When decaying, biomass requires oxygen. In an environment where growth is restricted by access to P, like in most freshwaters, 1 molecule of P creates 106 molecules of biomass C. 1 kg of BOD will therefore correspond to 0.00737 kg P in terms of DO depletion. 0.00737 is the ratio of the mass of 1 molecule of P relative to the 127.25 molecules of O_2 that is needed to oxidize 106 molecules of C, 16 molecules of N and one molecule of P. So, the contribution of 1 kg of N-tot to eutrophicated freshwater depends on both P and BOD emissions.

Total global BOD emission to freshwater estimated from country specific World Bank data for 2004–2007 is estimated to be 1.6E+10 kg. This estimation is based on an assumption that per capita emissions from the countries not covered by World Bank statistics (46% of the population) are the same as those covered. Global P-tot emissions to freshwater is estimated to be 3.8E+10 kg/yr[3].

Therefore, the contribution from 1 kg P-tot to DO depletion in freshwaters is 1/(1.6E+10*0.00737 + 3.8E+10) = 2.62E–11.

The environmental impact factor then becomes 4.97E+08*2.62E–11 = 1.30E–02 kg fish/kg P-tot.

The model builds on several assumptions: (1) Extrapolation of DO levels to waters not monitored, (2) independency between temperature and DO levels, (3) a linear relation between DO deficiency and decreased growth rate and (4) average production capacity of 1000 kg/km^2 (which is relevant for temperate regions). Therefore, a standard deviation of a factor of 2 is assumed.

10.4.3.2 Decreased biodiversity due to oxygen deficiency

According to IUCN red-list (http://www.iucnredlist.org/search, accessed on 15 October 2014), 0.007 of red-listed species are threatened by eutrophication. The contribution from P-tot is the same as in 10.4.3.1, 2.62E–11 per kg P-tot. The environmental impact factor is 0.007*2.62E–11= 1.83E–13 shares of threat to biodiversity per kg of N-tot.

It is unclear how well the data reported to IUCN cover the real situation. An uncertainty of a factor of 3 is assumed.

10.4.4 Summary and monetary valuation of P-tot emissions to freshwater

Environmental impact factors and monetary values for emissions of P-tot to freshwaters are shown in Table 10.3. As for N-tot, the impact value is moderate, and emissions of P-tot will not be a significant aspect unless it is central for the product or service, like in wastewater treatment.

10.5 As to Freshwater

10.5.1 Specification of the emission

The emission valued is emissions of arsenic to freshwater anywhere in the world in 2015.

10.5.2 Impacts pathways

Arsenic in drinking water and food cause bladder cancer, skin cancer and excess mortality in cardiovascular diseases.

10.5.3 Environmental impact factors

10.5.3.1 YLL via bladder cancer

The global yearly run-off of water is 24000 km^3 [6]. A person drinks about 2 l/day water as an average. The total intake of water from the global population is thus 2E–12*365*7.2E+09 = 5.256 km^3/year. Of 1 kg of emitted metal, 5.256/24000 = 0.000219 kg or 219 mg will enter humans through drinking water per year. This corresponds to a daily individual dose of 219/(365*7.2E+09)*1000 = 8.33E–08 μg/day.

Table 10.3. Summary of environmental impact factors and monetary impact values of emission of 1 kg P-tot to freshwater.

Environmental good	Impact indicator	Unit	Pathway	Environmental impact factor	Uncertainty	Indicator value ($/unit)	Impact value ($/kg P-tot)
fish	production capacity	kg	oxygen deficiency	1.30E–02	2	2	3.15E–02
biodiversity	share of threat to redlisted species	dimensionless	oxygen deficiency	1.83E–13	3	7.61E+10	1.40E–02
						sum	4.55E–02

The risk for bladder cancer is increased by 0.01%/µgAs/day (EU project ESPREME) and the global mortality due to bladder cancer is 1.7E+05 cases per year [7]. Thus, the extent of bladder cancers caused by 1 kg of As is 8.33E–08*0.0001*1.7E+05 =1.42E–06.

The marginal reduction of life expectancy is estimated as 3.7 years based on WHO estimates of global YLL/years for different age groups and an average life expectancy of 70 years.

Thus, the average environmental impact factor is1.42E–06*3.7 =2.9E–9 YLL/kg As.

In the model it was assumed that no purification of water was made. Other intake pathways, e.g., via food were not modeled. This introduces a large uncertainty. A factor of 10 is assumed.

10.5.3.2 YLL via cardio-vascular diseases

The global yearly run-off of water is 24000 km^3 [6]. A person drinks about 2 l/day water as an average. The total intake of water from the global population is thus 2E–12*365*7.2E+09 = 5.256 km^3/year. Of 1 kg of emitted metal, 5.256/24000 = 0.000219 kg or 219 mg will enter humans through drinking water per year. This corresponds to an average daily individual dose of 219/(365*7.2E+09)*1000 = 8.33E–08 µg/day.

There is a 0.0025% absolute risk/µg/day to die from CVD due to As exposure [8]. The marginal reduction of life expectancy is estimated to be 5.0 years based on WHO estimates of global YLL for different age groups and an average life expectancy of 70 years.

Thus, the average YLL/kg As is 0.000025*8.33E–08*7.2E+09*5 = 0.075 YLL/kg As.

In the model it was assumed that no purification of water was made. Other intake pathways, e.g., via food were not modeled. This introduces a large uncertainty. A factor of 10 is assumed.

10.5.3.3 YLL via direct exposure and skin cancer

The global yearly run-off of water is 24000 km^3 [6]. A person drinks about 2 l/day water as an average. The total intake of water from the global population is thus 2E–12*365*7.2E+09 = 5.256 km^3/year. Of 1 kg of emitted metal, 5.256/24000 = 0.000219 kg or 219 mg will enter humans through drinking water per year. This corresponds to an average daily individual dose of 219/(365*7.2E+09)*1000 = 8.33E–08 µg/day.

The relative risk for skin cancer including malignant melanoma is increased by 0.002%/µg/day [8] and the global average incidence is 0.00093 (49100 deaths out of total 52700000) [7]. Thus, there is an absolute risk of 2E–05*9.3E-4 = 1.86E-8 per ug/day lifetime exposure.

The marginal reduction of life expectancy from melanoma and skin cancer is estimated to be 7.4 years based on WHO estimates of global YLL/ years for different age groups and an average life expectancy of 70 years.

Mortality in skin cancer is low, about 12% for melanoma and 0.2% for SCC and BCC. If applying the ratio of predicted cancer incidences for melanoma vs. SCC + BCC, the US [9] which estimate the melanoma incidence to 2.7% of all cases, the global mortality rate until 2100 will be 0.12*0.027 + 0.002*0.973 = 0.00518.

Thus, the average YLL is 0.0000000186*0.0000000833*7200000000*0.00518*7.4 = 4.28E–7 YLL/kg As.

In the model it was assumed that no purification of water was made. Other intake pathways, e.g., via food, were not modeled. This introduces a large uncertainty. A factor of 10 is assumed.

10.5.4 Summary and monetary valuation of As emissions to freshwater

A summary of environmental impact factors and monetary impact values are shown in Table 10.4. It is notable that the impacts via increased cardiovascular diseases is so large compared to the other types of impacts.

10.6 Cd to Freshwater

10.6.1 Specification of the emission

Cd is emitted to freshwater through drainage from fertilizers and from specific industrial activities.

The emissions valued here is emissions of Cd in trace amounts anywhere in the world 2015.

10.6.2 Impacts pathways

Cd causes osteoporosis and renal dysfunction through ingestion. Here we only follow the pathways via direct intake via drinking water.

10.6.3 Environmental impact factors

10.6.3.1 Cases of osteoporosis via ingestion

The global yearly run-off of water is 24000 km^3 [6]. A person drinks about 2 l/day water as an average. The total intake of water from the global population is thus 2E–12*365*7.2E+09 = 5.256 km^3/year. Of 1 kg of emitted metal, 5.256/24000 = 0.000219 kg or 219 mg will enter humans through

Table 10.4. Summary of environmental impact factors and monetary impact values of emission of 1 kg As to freshwater.

Environmental good	Impact indicator	Unit	Pathway	Environmental impact factor	Uncertainty	Indicator value ($/unit)	Impact value ($/kg As)
human health	YLL	personyears	bladder cancer	2.90E–09	10	107,067	3.10E–04
human health	YLL	personyears	cardiovascular diseases	0.075	10	107,067	8.03E+03
human health	YLL	personyears	skin cancer	4.28E–07	10	107,067	4.58E–02
						sum	8.03E+03

drinking water per year. This corresponds to an average daily individual dose of 219/(365*7.2E+09)*1000 = 8.33E–08 µg Cd/day and kg Cd.

There is a 0.8% relative risk/µg/day to develop osteoporosis due to Cd exposure [8]. Osteoporosis is not monitored in the same way as other diseases, and no statistics on its prevalence have been found. According to Ross [10], there is a 50% risk of developing osteoporosis sometime in life for women and a 30% risk for men. Osteoporosis is mainly occurring in the elderly population. If the average 40% is used and the elderly 10% of the population suffers from osteoporosis, the absolute risk is 0.8*0.4*0.1 = 0.032%.

Thus, the average cases/kg Cd is 0.00032*8.33E–08*7.2E+09 = 1.91E–1.

In the model it was assumed that no purification of water was made. Other intake pathways, e.g., via food were not modeled. This introduces a large uncertainty. A factor of 10 is assumed.

10.6.3.2 YLD of renal dysfunction via ingestion

The global yearly run-off of water is 24000 km^3[6]. A person drinks about 2 l/day water as an average. The total intake of water from the global population is thus 2E–12*365*7.2E+09 = 5.256 km^3/year. Of 1 kg of emitted metal, 5.256/24000 = 0.000219 kg or 219 mg will enter humans through drinking water per year. This corresponds to an average daily individual dose of 219/(365*7.2E+09)*1000 = 8.33E–08 µg Cd/day and kg Cd.

There is a 0.04% relative risk/µg/day to develop renal dysfunction due to Cd exposure [8]. The global burden of kidney diseases is 4.018E+06 YLD 2010 [11]. The environmental impact factor is thus 0.0004*8.33E-08*4.018E+06 = 1.34E–04 YLD/kg Cd.

In the model it was assumed that no purification of water was made. Other intake pathways, e.g., via food were not modeled. This introduces a large uncertainty. A factor of 10 is assumed.

10.6.4 Summary and monetary valuation of Cd emissions to freshwater

A summary of environmental impact factors and monetary impact values are shown in Table 10.5.

The monetary impact value for osteoporosis is quite high. It is clear that it is a very large handicap and reduces the working capacity considerably. But it is unclear how prevalent the severe cases are.

Table 10.5. Summary of environmental impact factors and monetary impact values of emission of 1 kg Cd to freshwater.

Environmental good	Impact indicator	Unit	Pathway	Environmental impact factor	Uncertainty	Indicator value ($/unit)	Impact value ($/kg Cd)
human health	cases	personyears	osteoporosis	1.91E–01	10	137,045	2.62E+04
human health	YLD	personyears	renal dysfunction	1.34E–04	10	107,067	1.43E+01
						sum	2.62E+04

10.7 Hg to Water (and air)

Mercury is to a large extent moving around in the environment. It transforms from one chemical state to another and is transported globally in air as Hg^0. This means that the impact values for emission to water can be seen as approximately independent of where the emission occurs and to what media.

10.7.1 Specification of the emission

The emission valued is any type of Hg emissions anywhere in the world 2015.

10.7.2 Impacts pathways

Some groups of the population in North America, Europe and New Zealand eating much locally caught fish tend to get high mercury concentrations in body tissue. This may lead to various health issues but the one of most concern is mental retardation of children due to prenatal exposure [12].

10.7.3 Environmental impact factors

10.7.3.1 Personyears of mild intellectual disability via ingestion

In a New Zealand study, 1000 out of 11000 new mothers had consumed fish more than three times a week. 73 of these had hair mercury levels above 6 mg/kg. 50% of the high mercury level children had abnormal or questionable test results in a Denver Development Screening Test, whereas only 17% of the reference children had such results. This indicates that 0.2% of a "fish eating" population like New Zealand is affected. Globally the "fish eating" population is in the order of 200 million. 0.2% of these are 400000.

In a USEPA study to the Congress 1997, 166000 pregnant women were estimated to be eating fish above 100 g/day. This corresponds to 0.07% of the population. Upscaled to the global population, 4.8 million people/year will be eating fish above 100 g/day. If the same frequency of high mercury levels is assumed as in New Zealand (73/1000), we obtain a figure of 0.073*4.8 million = 345000 personyears per year.

In an Iraqi study cited by USEPA, severe neurogical effects is observed first at hair concentrations above 10–50 mg/kg. The effect corresponding to DALY classification of mild mental retardations is therefore estimated to 1000000 personyears per year.

The global anthropogenic emission of Hg to air is estimated to be 1960 tonnes and 1000 tonnes to water in 2010 (UNEP 2010). The natural emission is around 670 tonnes, and there is a substantial reemission of mercury from old depositions, approximately 4000 tonnes. Totally the emissions contributing to the population exposure via fish is thus 7630 ton.

The environmental impact factor is then 1E+06/7.63E+06 = 0.131 personyears/kg Hg.

The uncertainty is estimated to a factor of 3, mainly because of uncertain extent and severity of the sequalae.

10.7.4 Monetary valuation

The monetary value for mild intellectual disability is 3 319 $/personyear. The monetary impact value for an emission of Hg to water is 0.131*3319 = 435 $/kg Hg.

References

1. USGS, *Five-day biochemical oxygen demand: U.S. Geological Survey Techniques of Water-Resources Investigations,*. 2003, USGS.
2. *GEMSTAT*. 2014, UN.
3. Liu, Y., et al., Global phosphorus flows and environmental impacts from a consumption perspective. *Journal of Industrial Ecology*, 2008. **12**(2): 229–247.
4. Diaz, R.J. and R. Rosenberg, Spreading dead zones and consequences for marine ecosystems. *Science*, 2008. **321**(5891): 926–929.
5. Galloway, J.N. et al., Nitrogen cycles: past, present, and future. *Biogeochemistry*, 2004. **70**(2): 153–226.
6. Gerten, D. et al., Terrestrial vegetation and water balance—hydrological evaluation of a dynamic global vegetation model. *Journal of Hydrology*, 2004. **286**: 249–270.
7. Lozano, R. et al., *Global and regional mortality from 235 causes of death for 20 age groups in 1990 and 2010: a systematic analysis for the Global Burden of Disease Study 2010*. The Lancet, 2012. **380**(9859): 2095–2128.
8. Friedrich, R., *ESPREME, Estimation of willingness-to-pay to reduce risks of exposure to heavy metals and cost-benefits analysis for reducing heavy metals occurrence in Europe*, in *Activity report*. 2007, EU 6th framework program. p. 58.
9. Consulting, I., *Human Health Benefits of Straospheric Ozone Protection*. 2006, U.S. Environmental Protection Agency: Washington, DC 20460.
10. Ross, P.D., Osteoporosis. Frequency, consequences, and risk factors. *Arch Intern Med.*, 1996 Jul 8; 156(13): 1996. **156**(13): 1399–411.
11. Vos, T. et al., Years lived with disability (YLDs) for 1160 sequelae of 289 diseases and injuries 1990–2010: a systematic analysis for the Global Burden of Disease Study 2010. *The Lancet*, 2012. **380**(9859): 2163–2196.
12. Kjellström, T. et al., *Physical and Mental Development of Children with Prenatal Exposure to Mercury from Fish*. 1988, National Swedish Environmental Protection Board.

11

Use of Pesticides

Environmental impacts from the use of pesticides have a significant role in early environmental awareness. Rachel Carson's *Silent Spring* put the light on impacts on birds from [1]. It was estimated that 1992 about 800 different agents were used [2]. DDT and other pesticides and insecticides have been banned in many countries. Between 1993 and 2008, 704 active substances were banned in the EU [3]. The use of pesticides per area have decreased. In the 1960s as much as 1 kg was used per hectare, while nowadays only 10 g per hectare of an active substance may be used [4]. In USA and in the EU, there are special registration processes for controlling which pesticides may be used. The registration process assesses cost and benefits of the pesticide and is quite expensive. This means that risks and benefits of pesticides are relatively well known, although large uncertainties remain on the use of individual pesticides and their epidemiology. FAO collects and reports data on the use of different groups of pesticides in their database FAOSTAT, but there is a lack of corresponding data on health effects. Only data on total impacts are available. For individual pesticides, WHO publishes extensive information on the toxicology and properties of individual pesticides [5].

11.1 Specification of the Elementary Flow

Use of pesticides is not an emission in its normal meaning. It is an elementary flow that involves emissions to air, water and soil, but also impacts from the life cycle of the pesticides. In a CBA or LCA context, specific knowledge of how pesticides were produced or applied is seldom available. When buying agricultural products in the market, at most, the type of pesticides and amounts used in its production may be known.

The elementary flow is application of pesticides on soil and vegetation anywhere in the world in 2015.

11.2 Impact Pathways

Environmental impacts from the use of pesticides may be defined as those not affecting the target organism(s). WHO publishes monographs on individual pesticides, where the present knowledge of environmental impacts from the *use phase* are summarized. Practically no information is given about the other phases of their life cycles. But even with the information given about the use phase, the impacts and related pathways are many and seldom quantified.

Table 11.1 shows the impacts that have been possible to quantify.

They are health impacts, impacts on biodiversity and depletion of natural resources. Health impacts are diverse. Different target organs are affected, but statistics on impacts on individual organs are rare. Health impacts are both acute and chronic. In the publication on global burden of

Table 11.1. Environmental impacts and pathways at pesticide use.

Environmental good	Impact indicator	Unit	Pathway
human health	YLL	personyears	direct and indirect exposure
human health	YLL	personyears	CO_2-emission, climate change
human health	YLD	personyears	direct and indirect exposure
human health	working capacity	personhours	CO_2-emision, climate change
biodiversity	share of threat to redlisted species	dimensionless	direct and indirect exposure
natural gas resources	amount extracted	kg	used in production
oil resources	amount extracted	kg	used in production
coal resources	amount extracted	kg	used in production
As-ore	amount extracted	kg	used in production
Cu-ore	amount extracted	kg	used in production
Hg-ore	amount extracted	kg	used in production
Pb-ore	amount extracted	kg	used in production
P-resources	amount extracted	kg	used in production
S-resources	amount extracted	kg	used in production
Zn-ore	amount extracted	kg	used in production

diseases [6] there is a category "poisoning" that would be relevant. Some pesticides cause cancer. USEPA classify 51 pesticides as carcinogenic. The IARC database classify 8 pesticides as carcinogenic [3]. Cancer effects have not been possible to model separately.

Some impacts are indirect and originate from the production and waste phases of the pesticide life cycle. There are a multitude of impacts from production and end-of-life phases, but here the pathways modelled are limited to impacts that are known to have high environmental costs, i.e., health effects of CO_2 and extraction of finite natural resources, i.e., natural gas (used for production of organic chemicals), arsenic, copper, fluorine, lead, mercury, phosphorus, sulphur and zinc. Bromine and chlorine are also used for pesticide production but are already today produced in a sustainable way and is not contributing to the impact cost.

11.3 Environmental Impact Factors

11.3.1 YLL from direct and indirect exposure

Estimations of the number of deaths due to pesticides vary. Miller estimates that there are 18000 deaths per year caused by pesticides [7]. Lozano et al. [8] report that there are totally 180400 deaths per year from poisoning and 883000 deaths from self-harm. Gunnel et al. [9] estimate that there are as much as 371 000 cases of suicides per year where pesticides are used. 2015, the total use of pesticides is around 4 million tonnes per year [10]. An average shortening of life expectancy of 35 years is assumed and the potency factor $P_i(1/LD_{50})$ of the substance, is used to allocate part of the total effect to the particular substance. As no data has been found for the amounts of individual pesticides used, it is assumed that the same amount is used of all pesticides. The YLL for 1 kg of the pesticide *i* is estimated as $18000*35/4E+09*1/n *P_i/P_{average}$, where n is the number of pesticides assumed to contribute to the impacts and $P_{average}$ is the average potency factor.

The uncertainty is large, partly because of lack of data on how much of each pesticide that is used and partly because of varying exposure between industrialized countries and developing countries. One may also ascribe different part of the suicide cases to pesticides. Gunnel et al. mention the fact that pesticides are easily available in society, and several cases of suicides committed on impulse, would not have happened without them. A factor 10 to 100 is assumed for the uncertainty. The factor 10 may be representative if the use is in the developing countries, where almost all impacts occur, and 100 if it is unknown where it is used.

11.3.2 YLL from CO_2 emissions during production

The LCA database Ecoinvent (www.ecoinvent.org) contains a dataset that was calculated from the arithmetic mean of all inputs and outputs of all 78 pesticides. The emission of CO_2 from production of 1 kg of this "average" pesticide was determined to 8.7 kg. 1 kg of CO_2 causes 1.88E–06 YLL (Table 6.1). The production of 1 kg of pesticide thus cause 8.7*1.88E–06 = 1.64E–05 YLL.

The uncertainty for the YLLs from 1 kg of CO_2 was estimated to a factor of 2. The variation in emissions of CO_2 for among specific pesticides is estimated to another factor of 2. The total uncertainty is therefore exp $\sqrt{(\ln 2)^2 + (\ln 2)^2} = 2.7$.

11.3.3 YLD from direct and indirect exposure

There are 417000 YLD due to accidental poisoning per year globally [11]. Most of this, 300000 YLD, are assumed to be caused by pesticides. The total use of pesticides is 4 million tonnes/year [10].

The potency factor $P_i(1/LD_{50})$ of the substance *i*, is used to allocate part of the total effect to the substancei. As no data has been found for the amounts of individual pesticides used, it is assumed that the same amount is used of all pesticides. The YLD for 1 kg of the pesticide *i* is estimated as 300000/4E+09*1/n *$P_i/P_{average}$, where n is the number of pesticides assumed to contribute to the impacts and $P_{average}$ is the average potency factor.

The uncertainty is large, partly because of lack of data on how much of each pesticide that is used and partly because of varying exposure between industrialized countries and developing countries. A factor 10 to 100 is assumed. The factor 10 may be representative if the use is in the developing countries, where almost all impacts occur, and 100 if it is unknown where it is used.

11.3.4 Decreased working capacity from CO_2 emissions during production

The LCA database Ecoinvent (www.ecoinvent.org) contains a dataset that was calculated from the arithmetic mean of all inputs and outputs of all 78 pesticides. The emission of CO_2 from production of 1 kg of this "average" pesticide was determined to be 8.7 kg. 1 kg of CO_2 causes 4.53E–03 personhours decreased working capacity (Table 6.1). The production of 1 kg of pesticide thus cause 8.7*4.53E–03 = 3.94E–02 personhours of decreased working capacity.

The uncertainty for the YLLs from 1 kg of CO_2 was estimated to a factor of 2. The variation in emissions of CO_2 for among specific pesticides is estimated to another factor of 2. The total uncertainty is therefore exp $\sqrt{(\ln 2)^2 + (\ln 2)^2} = 2.7$.

11.3.5 Decreased biodiversity

Pesticides are responsible for 0.8382% of the threat causes for red-listed species (IUCN 2014). The global use (= emission) of pesticides is 4 million tonnes/year. The potency factor of the substance is used to allocate part of the total effect to the particular substance.

The potency factor $P_i(1/LD_{50})$ of the substance *i*, is used to allocate part of the total effect to the substance *i*. As no data has been found for the amounts of individual pesticides used, it is assumed that the same amount is used of all pesticides. The share of threat to red-listed species for 1 kg of the pesticide *i* is estimated as 0.8382/4E+09*1/n *$P_i/P_{average}$, where n is the number of pesticides assumed to contribute to the impacts and $P_{average}$ is the average potency factor.

The uncertainty is large, partly because of lack of data on how much of each pesticide that is used and partly because of lack of knowledge on how potency relates to the redlisting. A factor of 100 is assumed.

11.3.6 Decreased natural resources

Production of pesticides requires use of several finite natural resources, such as fossil minerals, metal ores, P and S. No recycling of the resources exists, so there is a resource depletion equal to the amounts in the individual pesticide. The use of fossil resources is, as an average: 1.42 kg lignite, 0.95 kg hard coal, 2.67 m^3 natural gas and 0.8428 kg oil. The data for metal ores, P and S are presented in Table 11.2.

The uncertainty is relatively low. It is caused by resources lost in the production and not available in the pesticides. These losses are estimated to at most 20%.

11.4 Summary of Impact Factors and Monetary Values for Use of Pesticides

For about 9% of the pesticides in Table 11.2, more than half of the impact cost is due to use of scarce elements such as As, Cu, F, I, Hg, Sn, Pb, Tl and Zn.

Table 11.2. Impact factors and monetary values of environmental impacts from use of pesticides.The impact factors at production is not shown in the table. The impact costs at production, 8.22$/kg, is added directly to the last column.

Substance name	CASRN	Potency factor	YLL, acute tox	Poisoning personyear	Resources (kg)	Biodiversity	Impact cost ($/kg)
2,3,6-TBA	000050-31-7	6.67E–01	2.05E–06	1.95E–07		2.73E–14	8.46E+00
2,4-D	000094-75-7	2.67E+00	8.21E–06	7.82E–07		1.09E–13	9.16E+00
2,4-DB	000094-82-6	1.43E+00	4.40E–06	4.19E–07		5.85E–14	8.73E+00
2-Napthyloxyacetic acid	000120-23-0	1.67E+00	5.13E–06	4.89E–07		6.82E–14	8.81E+00
3-Chloro-1,2-propanediol	000096-24-2	8.93E+00	2.75E–05	2.62E–06		3.66E–13	1.14E+01
4-CPA	000122-88-3	1.18E+00	3.62E–06	3.45E–07		4.82E–14	8.64E+00
Acephate	030560-19-1	1.06E+00	3.26E–06	3.10E–07	P: 0.169, S: 0.175	4.33E–14	1.00E+01
Acifluorfen	050594-66-6	7.30E–01	2.25E–06	2.14E–07	F: 0.157	2.99E–14	9.43E+00
Acrolein	000107-02-8	3.45E+01	1.06E–04	1.01E–05		1.41E–12	2.03E+01
Alachlor	015972-60-8	1.08E+00	3.31E–06	3.15E–07		4.40E–14	8.60E+00
Alanycarb	083130-01-2	3.03E+00	9.33E–06	8.88E–07	S: 0.16	1.24E–13	9.32E+00
Aldicarb	000116-06-3	1.08E+03	3.31E–03	3.15E–04	S: 0.168	4.40E–11	3.86E+02
Allethrin	000584-79-2	1.46E+00	4.49E–06	4.28E–07		5.98E–14	8.74E+00
Allyl alcohol	000107-18-6	1.56E+01	4.81E–05	4.58E–06		6.40E–13	1.37E+01
Alpha-cypermethrin	067375-30-8	1.27E+01	3.90E–05	3.71E–06		5.18E–13	1.27E+01
Ametryn	000834-12-8	9.09E+00	2.80E–05	2.66E–06		3.72E–13	1.14E+01
Amitraz	033089-61-1	1.25E+00	3.85E–06	3.66E–07		5.12E–14	8.66E+00
Anilofos	064249-01-0	2.12E+00	6.52E–06	6.21E–07	P: 0.0843, S: 0.174	8.68E–14	9.70E+00

Table 11.2 contd. ...

... Table 11.2 contd.

Substance name	CASRN	Potency factor	YLL, acute tox	Poisoning personyear	Resources (kg)	Biodiversity	Impact cost ($/kg)
Azaconazole	060207-31-0	3.25E+00	9.99E–06	9.52E–07		1.33E–13	9.37E+00
Azamethiphos	035575-96-3	9.90E–01	3.05E–06	2.90E–07	P: 0.0955, S: 0.0986	4.05E–14	9.38E+00
Azinphos-ethyl	002642-71-9	8.33E+01	2.56E–04	2.44E–05	P: 0.0898, S: 0.185	3.41E–12	3.83E+01
Azinphos-methyl	000086-50-0	6.25E+01	1.92E–04	1.83E–05	P: 0.0977, S: 0.202	2.56E–12	3.10E+01
Azocyclotin	041083-11-8	1.25E+01	3.85E–05	3.66E–06	Sn: 0.272	5.12E–13	2.03E+02
Bendiocarb	022781-23-3	1.82E+01	5.60E–05	5.33E–06		7.45E–13	1.46E+01
Benfuracarb	082560-54-1	4.88E+00	1.50E–05	1.43E–06	S: 0.078	2.00E–13	9.95E+00
Bensulide	000741-58-2	3.70E+00	1.14E–05	1.09E–06	P: 0.0780, S: 0.242	1.52E–13	1.02E+01
Bensultap	017606-31-4	9.09E–01	2.80E–06	2.66E–07	S: 0.297	3.72E–14	8.60E+00
Bentazone	025057-89-0	9.09E–01	2.80E–06	2.66E–07	S: 0.133	3.72E–14	8.57E+00
Beta-cyfluthrin	068359-37-5	9.09E+01	2.80E–04	2.66E–05	F: 0.0438	3.72E–12	4.04E+01
Bifenthrin	082657–04-3	1.82E+01	5.60E–05	5.33E–06	F: 0.135	7.45E–13	1.54E+01
Bilanafos	071048-99-2	3.73E+00	1.15E–05	1.09E–06	P: 0.096	1.53E–13	1.03E+01
Bioallethrin	000584-79-2	1.43E+00	4.40E–06	4.19E–07		5.85E–14	8.73E+00
Blasticidin-S	002079-00-7	6.25E+01	1.92E–04	1.83E–05		2.56E–12	3.02E+01
Brodifacoum	056073-10-0	3.33E+03	1.03E–02	9.77E–04		1.36E–10	1.18E+03
Bromadiolone	028772-56-7	8.93E+02	2.75E–03	2.62E–04		3.66E–11	3.22E+02
Bromethalin	063333-35-7	5.00E+02	1.54E–03	1.47E–04		2.05E–11	1.84E+02
Bromoxynil	001689-84-5	5.26E+00	1.62E–05	1.54E–06		2.16E–13	1.01E+01
Bromuconazole	116255-48-2	2.74E+00	8.43E–06	8.03E–07		1.12E–13	9.19E+00

Bronopol	000052-51-7	3.94E+00	1.21E–05	1.15E–06		1.61E–13	9.61E+00
Butamifos	036335-67-8	1.59E+00	4.89E–06	4.65E–07	P: 0.0934, S: 0.0964	6.50E–14	9.57E+00
Butocarboxim	034681-10-2	6.33E+00	1.95E–05	1.86E–06	S: 0.168	2.59E–13	1.05E+01
Butoxycarboxim	034681-23-7	3.47E+00	1.07E–05	1.02E–06	S: 0.144	1.42E–13	9.47E+00
Butralin	033629-47-9	9.53E–01	2.93E–06	2.79E–07		3.90E–14	8.56E+00
Butroxydim	138164-12-2	6.12E–01	1.88E–06	1.79E–07		2.50E–14	8.44E+00
Butylamine	013952-84-6	2.63E+00	8.10E–06	7.71E–07		1.08E–13	9.15E+00
Cadusafos	095465-99-9	2.70E+01	8.32E–05	7.92E–06	P: 0.115, S: 0.237	1.11E–12	1.87E+01
Calcium arsenate	007778-44-1	5.00E+01	1.54E–04	1.47E–05	As: 0.188	2.05E–12	7.49E+02
Calcium cyanide	000592-01-8	2.56E+01	7.89E–05	7.52E–06		1.05E–12	1.72E+01
Captafol	002425-06-1	2.00E–01	6.16E–07	5.86E–08	S: 0.0917	8.19E–15	8.31E+00
Carbaryl	000063-25-2	3.33E+00	1.03E–05	9.77E–07		1.36E–13	9.40E+00
Carbofuran	001563-66-2	1.25E+02	3.85E–04	3.66E–05		5.12E–12	5.22E+01
Carbosulfan	055285-14-8	4.00E+00	1.23E–05	1.17E–06	S: 0.0841	1.64E–13	9.65E+00
Cartap	015263-53-3	3.08E+00	9.47E–06	9.02E–07	S: 0.270	1.26E–13	9.36E+00
Chloralose	015879-93-3	2.50E+00	7.69E–06	7.33E–07		1.02E–13	9.10E+00
Chlordane	000057-74-9	2.17E+00	6.69E–06	6.37E–07		8.90E–14	8.99E+00
Chlorethoxyfos	054593-83-8	5.56E+02	1.71E–03	1.63E–04	P: 0.0923, S: 0.0952	2.27E–11	2.04E+02
Chlorfenapyr	122453-73-0	2.27E+00	6.98E–06	6.65E–07	F: 14.0	9.29E–14	9.86E+00
Chlorfenvinphos	000470-90-6	3.23E+01	9.93E–05	9.46E–06	P: 8.62	1.32E–12	2.03E+01

Table 11.2 contd. ...

... Table 11.2 contd.

Substance name	CASRN	Potency factor	YLL, acute tox	Poisoning personyear	Resources (kg)	Biodiversity	Impact cost ($/kg)
Chlormephos	024934-91-6	1.43E+02	4.40E–04	4.19E–05	P: 13.2, S: 27.3	5.85E–12	5.96E+01
Chlormequat (chloride)	000999-81-5	1.49E+00	4.59E–06	4.37E–07		6.11E–14	8.75E+00
Chloroacetic acid	000079-11-8	1.54E+00	4.74E–06	4.51E–07		6.30E–14	8.77E+00
Chlorophacinone	003691-35-8	3.23E+02	9.93E–04	9.46E–05		1.32E–11	1.22E+02
Chlorphonium chloride	000115-78-6	5.62E+00	1.73E–05	1.65E–06	P: 0.779	2.30E–13	1.08E+01
Chlorpyrifos	002921-88-2	7.41E+00	2.28E–05	2.17E–06	P: 0.0882, S: 0.0912	3.03E–13	1.16E+01
Clomazone	081777-89-1	7.30E–01	2.25E–06	2.14E–07		2.99E–14	8.48E+00
Copper hydroxide	020427-59-2	1.00E+00	3.08E–06	2.93E–07	Cu: 0.651	4.09E–14	1.02E+02
Copper oxychloride	001332-40-7	6.94E–01	2.14E–06	2.04E–07	Cu: 0.596	2.84E–14	9.44E+01
Copper sulfate	007758-98-7	3.33E+00	1.03E–05	9.77E–07	Cu: 0.398	1.36E–13	6.68E+01
Coumaphos	000056-72-4	1.41E+02	4.33E–04	4.13E–05	P: 0.0854, S: 0.0881	5.77E–12	5.84E+01
Coumatetralyl	005836-29-3	6.25E+01	1.92E–04	1.83E–05		2.56E–12	3.02E+01
Cuprous oxide	001317-39-1	2.13E+00	6.55E–06	6.24E–07	Cu: 0.888	8.71E–14	1.37E+02
Cyanazine	021725-46-2	3.47E+00	1.07E–05	1.02E–06		1.42E–13	9.45E+00
Cyanophos	002636-26-2	1.64E+00	5.05E–06	4.81E–07	P: 0.128, S: 0.132	6.71E–14	9.87E+00
Cyfluthrin	068359-37-5	6.67E+01	2.05E–04	1.95E–05	F: 0.0438	2.73E–12	3.19E+01
Cyhalothrin	068085-85-8	6.94E+00	2.14E–05	2.04E–06	F: 0.127	2.84E–13	1.14E+01
Cyhexatin	013121-70-5	3.77E+00	1.16E–05	1.11E–06	Sn: 0.308	1.55E–13	2.25E+02
Cymoxanil	057966-95-7	8.36E–01	2.57E–06	2.45E–07		3.42E–14	8.52E+00
Cypermethrin	052315-07-8	4.00E+00	1.23E–05	1.17E–06		1.64E–13	9.63E+00

Cyphenothrin [(1R)-isomers]	039515-40-7	3.14E+00	9.68E–06	9.22E–07		1.29E–13	9.33E+00
Cyproconazole	094361-06-5	9.80E–01	3.02E–06	2.87E–07		4.01E–14	8.57E+00
Dazomet	000533-74-4	1.56E+00	4.81E–06	4.58E–07	S: 0.395	6.40E–14	8.85E+00
DDT	000050-29-3	8.85E+00	2.72E–05	2.59E–06		3.62E–13	1.13E+01
Deltamethrin	052918-63-5	7.41E+00	2.28E–05	2.17E–06		3.03E–13	1.08E+01
Demeton-S-methyl	000919-86-8	2.50E+01	7.69E–05	7.33E–06	P: 0.135, S: 0.278	1.02E–12	1.82E+01
Diazinon	000333-41-5	3.33E+00	1.03E–05	9.77E–07	P: 0.102, S: 0.0792	1.36E–13	1.02E+01
Dicamba	001918-00-9	5.86E–01	1.80E–06	1.72E–07		2.40E–14	8.43E+00
Dichlorobenzene	000106-46-7	3.64E–01	1.12E–06	1.07E–07		1.49E–14	8.35E+00
Dichlorophen	000097-23-4	8.00E–01	2.46E–06	2.34E–07		3.28E–14	8.51E+00
Dichlorprop	007547-66-2	1.25E+00	3.85E–06	3.66E–07		5.12E–14	8.66E+00
Dichlorvos	000062-73-7	1.79E+01	5.50E–05	5.23E–06	P: 0.14	7.31E–13	1.57E+01
Diclofop	040483-25-2	1.77E+00	5.45E–06	5.19E–07		7.25E–14	8.85E+00
Dicofol	000115-32-2	1.45E+00	4.46E–06	4.25E–07		5.93E–14	8.73E+00
Dicrotophos	000141-66-2	4.55E+01	1.40E–04	1.33E–05	P: 0.131	1.86E–12	2.53E+01
Difenacoum	056073-07-5	5.56E+02	1.71E–03	1.63E–04		2.27E–11	2.03E+02
Difenoconazole	119446-68-3	6.88E–01	2.12E–06	2.02E–07		2.82E–14	8.47E+00
Difenzoquat	043222-48-6	2.13E+00	6.55E–06	6.24E–07		8.71E–14	8.97E+00
Difethialone	104653-34-1	1.79E+03	5.50E–03	5.23E–04	S: 0.0593	7.31E–11	6.36E+02
Dimepiperate	061432-55-1	1.06E+00	3.25E–06	3.10E–07	S: 0.122	4.33E–14	8.62E+00

Table 11.2 contd. …

… Table 11.2 contd.

Substance name	CASRN	Potency factor	YLL, acute tox	Poisoning personyear	Resources (kg)	Biodiversity	Impact cost ($/kg)
Dimethachlor	050563-36-5	6.25E–01	1.92E–06	1.83E–07		2.56E–14	8.44E+00
Dimethenamid	087674-68-8	2.70E+00	8.30E–06	7.90E–07	S: 0.116	1.10E–13	9.20E+00
Dimethipin	055290-64-7	8.47E–01	2.61E–06	2.48E–07	S: 0.305	3.47E–14	8.58E+00
Dimethoate	000060-51-5	6.67E+00	2.05E–05	1.95E–06	P: 0.135, S: 0.279	2.73E–13	1.17E+01
Dimethylarsinic acid	000075-60-5	7.41E–01	2.28E–06	2.17E–07	As: 0.543	3.03E–14	2.09E+03
Diniconazole	083657-24-3	1.56E+00	4.82E–06	4.59E–07		6.41E–14	8.77E+00
Dinobuton	000973-21-7	7.14E+00	2.20E–05	2.09E–06		2.92E–13	1.07E+01
Dinocap	039300-45-3	1.02E+00	3.14E–06	2.99E–07		4.18E–14	8.58E+00
Dinoterb	001420-07-1	4.00E+01	1.23E–04	1.17E–05		1.64E–12	2.23E+01
Diphacinone	000082-66-6	4.35E+02	1.34E–03	1.27E–04		1.78E–11	1.61E+02
Diphenamid	000957-51-7	1.03E+00	3.17E–06	3.02E–07		4.22E–14	8.59E+00
Diquat	002764-72-9	4.33E+00	1.33E–05	1.27E–06		1.77E–13	9.75E+00
Disulfoton	000298-04-4	3.85E+02	1.18E–03	1.13E–04	P: 0.11.3 S: 0.350	1.57E–11	1.44E+02
Dithianon	003347-22-6	1.56E+00	4.81E–06	4.58E–07	S: 0.216	6.40E–14	8.82E+00
DNOC	000534-52-1	4.00E+01	1.23E–04	1.17E–05		1.64E–12	2.23E+01
Dodine	002439-10-3	1.00E+00	3.08E–06	2.93E–07		4.09E–14	8.58E+00
Edifenphos	017109-49-8	6.67E+00	2.05E–05	1.95E–06	P: 0.100, S: 0.206	2.73E–13	1.14E+01
Endosulfan	000115-29-7	1.25E+01	3.85E–05	3.66E–06	S: 0.0786	5.12E–13	1.26E+01
Endothal-sodium	000125-67-9	1.96E+01	6.03E–05	5.75E–06		8.03E–13	1.51E+01
EPN	002104-64-5	7.14E+01	2.20E–04	2.09E–05	P: 0.0960, S: 0.0991	2.92E–12	3.41E+01

EPTC	000759-94-4	6.05E–01	1.86E–06	1.77E–07	S: 0.169	2.48E–14	8.47E+00
Esfenvalerate	066230–04-4	1.15E+01	3.54E–05	3.37E–06		4.71E–13	1.23E+01
Ethiofencarb	029973-13-5	5.00E+00	1.54E–05	1.47E–06	S: 0.142	2.05E–13	1.00E+01
Ethion	000563-12-2	4.81E+00	1.48E–05	1.41E–06		1.97E–13	9.91E+00
Ethoprophos	013194-48-4	3.03E+01	9.33E–05	8.88E–06	P: 0.128, S: 0.264	1.24E–12	2.00E+01
Famphur	000052-85-7	2.08E+01	6.41E–05	6.11E–06	P: 0.0954, S: 0.197	8.53E–13	1.64E+01
Fenamiphos	022224-92-6	6.67E+01	2.05E–04	1.95E–05	P: 0.102, S: 0.106	2.73E–12	3.25E+01
Fenazaquin	120928–09-8	7.46E+00	2.30E–05	2.19E–06		3.06E–13	1.08E+01
Fenitrothion	000122-14-5	1.99E+00	6.12E–06	5.83E–07	P: 0.112, S: 0.116	8.14E–14	9.86E+00
Fenobucarb	003766-81-2	1.61E+00	4.96E–06	4.73E–07		6.60E–14	8.79E+00
Fenothiocarb	062850-32-2	8.70E–01	2.68E–06	2.55E–07	S: 0.126	3.56E–14	8.56E+00
Fenpropathrin	064257-84-7	1.52E+01	4.66E–05	4.44E–06		6.20E–13	1.36E+01
Fenpropidin	067306–00-7	8.77E–01	2.70E–06	2.57E–07		3.59E–14	8.53E+00
Fenpyroximate	134098-61-6	4.08E+00	1.26E–05	1.20E–06		1.67E–13	9.66E+00
Fenthion	000055-38-9	1.71E+00	5.25E–06	5.00E–07	P: 0.112, S: 0.230	6.99E–14	9.78E+00
Fentin acetate[(ISO)]	000900-95-8	8.00E+00	2.46E–05	2.34E–06	Sn: 0.290	3.28E–13	2.14E+02
Fentin hydroxide[(ISO)]	000076-87-9	9.26E+00	2.85E–05	2.71E–06	Sn: 0.322	3.79E–13	2.37E+02
Fenvalerate	051630-58-1	2.22E+00	6.84E–06	6.51E–07		9.10E–14	9.01E+00
Ferimzone	089269-64-7	1.38E+00	4.25E–06	4.04E–07		5.65E–14	8.71E+00
Fipronil	120068-37-3	1.09E+01	3.35E–05	3.19E–06	F: 0.261, S 0.0732	4.45E–13	1.36E+01

Table 11.2 contd. …

... Table 11.2 contd.

Substance name	CASRN	Potency factor	YLL, acute tox	Poisoning personyear	Resources (kg)	Biodiversity	Impact cost ($/kg)
Flocoumafen	090035-08-8	4.00E+03	1.23E–02	1.17E–03	F: 0.105	1.64E–10	1.41E+03
Fluchloralin	033245-39-5	6.45E–01	1.99E–06	1.89E–07	F: 0.161	2.64E–14	9.41E+00
Flucythrinate	070124-77-5	1.49E+01	4.59E–05	4.37E–06	F: 0.0843	6.11E–13	1.40E+01
Flufenacet	014245-95-8	1.67E+00	5.13E–06	4.89E–07	F: 0.209, S: 0.0882	6.82E–14	1.01E+01
Fluoroacetamide	000640-19-7	7.69E+01	2.37E–04	2.25E–05	F: 0.247	3.15E–12	3.67E+01
Fluoroglycofen	077501-60-1	6.45E–01	1.99E–06	1.89E–07	F: 0.0452	2.64E–14	8.72E+00
Flurprimidol	056425-91-3	1.41E+00	4.34E–06	4.13E–07	F: 0.183	5.78E–14	9.82E+00
Flusilazole	085509-19-9	1.49E+00	4.58E–06	4.36E–07	F: 0.121	6.09E–14	9.47E+00
Flutriafol	076674-21-0	8.77E–01	2.70E–06	2.57E–07	F: 0.126	3.59E–14	9.29E+00
Fluxofenim	088485-37-4	1.49E+00	4.59E–06	4.37E–07	F: 0.184	6.11E–14	9.85E+00
Fomesafen	072178–02-0	8.00E–01	2.46E–06	2.34E–07	F: 0.130, S: 0.0729	3.28E–14	9.30E+00
Formetanate	022259-30-9	4.76E+01	1.47E–04	1.40E–05		1.95E–12	2.50E+01
Fuberidazole	003878-19-1	2.98E+00	9.16E–06	8.72E–07		1.22E–13	9.27E+00
Furalaxyl	057646-30-7	1.06E+00	3.27E–06	3.12E–07		4.36E–14	8.60E+00
Furathiocarb	065907-30-4	2.38E+01	7.33E–05	6.98E–06	S: 0.0838	9.75E–13	1.66E+01
Gamma-HCH, Lindane	000058-89-9	1.14E+01	3.50E–05	3.33E–06		4.65E–13	1.22E+01
Glufosinate	053369–07-6	6.15E–01	1.89E–06	1.80E–07	P: 0.171	2.52E–14	9.85E+00
Guazatine	108173-90-6	4.35E+00	1.34E–05	1.27E–06		1.78E–13	9.75E+00
Haloxyfop	069806-34-4	3.33E+00	1.03E–05	9.77E–07	F: 0.157	1.36E–13	1.03E+01

HCH	000608-73-1	1.00E+01	3.08E–05	2.93E–06		4.09E–13	1.17E+01
Heptenophos	023560-59-0	1.04E+01	3.21E–05	3.05E–06	P: 0.124	4.27E–13	1.29E+01
Hexachlorobenzene	000118-74-1	1.00E–01	3.08E–07	2.93E–08		4.09E–15	8.26E+00
Hexazinone	051235-04-2	5.92E–01	1.82E–06	1.73E–07		2.42E–14	8.43E+00
Hydramethylnon	067485-29-4	8.33E–01	2.56E–06	2.44E–07	F: 0.231	3.41E–14	9.90E+00
Imazalil	035554-44-0	4.41E+00	1.36E–05	1.29E–06		1.80E–13	9.77E+00
Imidacloprid	138261-41-3	2.22E+00	6.84E–06	6.51E–07		9.10E–14	9.01E+00
Iminoctadine	013516-27-3	3.33E+00	1.03E–05	9.77E–07		1.36E–13	9.40E+00
Indoxacarb	173584-44-6	3.73E+00	1.15E–05	1.09E–06	F: 0.108	1.53E–13	1.02E+01
Ioxynil	001689-83-4	9.09E+00	2.80E–05	2.66E–06	I: 0.684	3.72E–13	3.19E+01
Ioxynil octanoate	003861-47-0	2.56E+00	7.89E–06	7.52E–07	I: 0.511	1.05E–13	2.44E+01
Iprobenfos	026087-47-8	1.67E+00	5.13E–06	4.89E–07	P: 0.108, S: 0.111	6.82E–14	9.72E+00
Isoprocarb	002631-40-5	2.48E+00	7.64E–06	7.27E–07		1.02E–13	9.10E+00
Isoprothiolane	050512-35-1	8.40E–01	2.59E–06	2.46E–07	S: 0.221	3.44E–14	8.56E+00
Isoproturon	034123-59-6	5.56E–01	1.71E–06	1.63E–07		2.27E–14	8.42E+00
Isouron	055861-78-4	1.59E+00	4.89E–06	4.65E–07		6.50E–14	8.78E+00
Isoxathion	018854-04-8	8.93E+00	2.75E–05	2.62E–06	P: 0.0990, S: 0.102	3.66E–13	1.22E+01
Lambda-cyhalothrin	091465-08-6	1.79E+01	5.50E–05	5.23E–06	F: 0.127	7.31E–13	1.53E+01
Lead arsenate	007784-40-9	1.00E+02	3.08E–04	2.93E–05	As: 0.216, Pb: 0.596	4.09E–12	1.08E+03
MCPA	000094-74-6	1.43E+00	4.40E–06	4.19E–07		5.85E–14	8.73E+00
MCPA-thioethyl	025319-90-8	1.27E+00	3.90E–06	3.71E–07	S: 0.131	5.18E–14	8.70E+00

Table 11.2 contd. ...

... Table 11.2 contd.

Substance name	CASRN	Potency factor	YLL, acute tox	Poisoning personyear	Resources (kg)	Biodiversity	Impact cost ($/kg)
MCPB	000094-81-5	1.47E+00	4.53E–06	4.31E–07		6.02E–14	8.74E+00
Mecarbam	002595-54-2	2.78E+01	8.55E–05	8.14E–06	P: 0.0942, S: 0.195	1.14E–12	1.88E+01
Mecoprop	007085-19-0	1.08E+00	3.31E–06	3.15E–07		4.40E–14	8.60E+00
Mecoprop-P	016484-77-8	9.52E–01	2.93E–06	2.79E–07		3.90E–14	8.56E+00
Mefluidide	053780-34-0	5.21E–01	1.60E–06	1.53E–07	F: 0.184, P: 0.103	2.13E–14	9.53E+00
Mepiquat	015302-91-7	6.71E–01	2.07E–06	1.97E–07		2.75E–14	8.46E+00
Mercuric chloride	007487-94-7	1.00E+03	3.08E–03	2.93E–04	Hg: 0.739	4.09E–11	6.39E+04
Mercuric oxide	021908-53-2	5.56E+01	1.71E–04	1.63E–05	Hg: 0.926	2.27E–12	7.96E+04
Mercurous chloride	010112-91-1	4.76E+00	1.47E–05	1.40E–06	Hg: 0.851	1.95E–13	7.32E+04
Metalaxyl	057837-19-1	1.49E+00	4.59E–06	4.37E–07		6.11E–14	8.75E+00
Metaldehyde	000108-62-3	4.41E+00	1.36E–05	1.29E–06		1.80E–13	9.77E+00
Metamitron	041394-05-2	8.45E–01	2.60E–06	2.48E–07		3.46E–14	8.52E+00
Metam-sodium [(ISO)]	000137-42-8	3.51E+00	1.08E–05	1.03E–06	S: 0.496	1.44E–13	9.56E+00
Metconazole	125116-23-6	1.52E+00	4.66E–06	4.44E–07		6.20E–14	8.76E+00
Methacrifos	062610-77-9	1.47E+00	4.54E–06	4.32E–07	P: 0.129, S: 0.133	6.04E–14	9.83E+00
Methamidophos	010265-92-6	3.33E+01	1.03E–04	9.77E–06	P: 0.220, S: 0.227	1.36E–12	2.18E+01
Methasulfocarb	066952-49-6	8.93E+00	2.75E–05	2.62E–06	S: 0.245	3.66E–13	1.14E+01
Methidathion	000950-37-8	4.00E+01	1.23E–04	1.17E–05	P:0.102, S: 0.318	1.64E–12	2.32E+01
Methiocarb	002032-65-7	5.00E+01	1.54E–04	1.47E–05	S: 0.284	2.05E–12	2.59E+01
Methomyl	016752-77-5	5.88E+01	1.81E–04	1.72E–05	S: 0.198	2.41E–12	2.89E+01

Methyl isothiocyanate	000556-61-6	1.39E+01	4.27E–05	4.07E–06	S: 0.438	5.69E–13	1.32E+01
Methylarsonic acid	000124-58-3	5.56E–01	1.71E–06	1.63E–07	As: 0.536	2.27E–14	2.07E+03
Metolcarb	001129-41-5	3.73E+00	1.15E–05	1.09E–06		1.53E–13	9.54E+00
Metribuzin	021087-64-9	3.11E+00	9.56E–06	9.10E–07	S: 0.150	1.27E–13	9.35E+00
Mevinphos	026718-65-0	2.70E+02	8.32E–04	7.92E–05	P: 0.138	1.11E–11	1.04E+02
Molinate	002212-67-1	1.39E+00	4.27E–06	4.07E–07	S: 0.171	5.69E–14	8.75E+00
Monocrotophos	006923-22-4	7.14E+01	2.20E–04	2.09E–05	P: 0.139	2.92E–12	3.45E+01
Myclobutanil	088671-89-0	6.25E–01	1.92E–06	1.83E–07		2.56E–14	8.44E+00
Nabam	000142-59-6	2.53E+00	7.79E–06	7.42E–07	S: 0.500	1.04E–13	9.21E+00
Naled	000300-76-5	2.33E+00	7.16E–06	6.82E–07	P: 0.814	9.52E–14	9.71E+00
Nicotine	000054-11-5	2.00E+01	6.16E–05	5.86E–06		8.19E–13	1.53E+01
Nitrapyrin	001929-82-4	9.33E–01	2.87E–06	2.73E–07		3.82E–14	8.55E+00
Nuarimol	063284-71-9	8.00E–01	2.46E–06	2.34E–07	F: 0.0603	3.28E–14	8.87E+00
Octhilinone	026530-20-1	6.80E–01	2.09E–06	1.99E–07	S: 0.150	2.79E–14	8.49E+00
Omethoate	001113-02-6	2.00E+01	6.16E–05	5.86E–06	P: 0.146, S: 0.150	8.19E–13	1.65E+01
Oxadixyl	077732-09-3	5.38E–01	1.65E–06	1.58E–07		2.20E–14	8.41E+00
Oxamyl	023135-22-0	1.67E+02	5.13E–04	4.89E–05	S: 0.146	6.82E–12	6.68E+01
Oxydemeton-methyl	000301-12-2	1.54E+01	4.74E–05	4.51E–06	P: 0.126, S: 0.260	6.30E–13	1.47E+01
Paclobutrazol	076738-62-0	7.69E–01	2.37E–06	2.25E–07		3.15E–14	8.50E+00
Paraquat	001910-42-5	6.67E+00	2.05E–05	1.95E–06		2.73E–13	1.06E+01
Parathion	000056-38-2	7.69E+01	2.37E–04	2.25E–05	P: 0.107, S: 0.110	3.15E–12	3.62E+01

Table 11.2 contd. …

... Table 11.2 contd.

Substance name	CASRN	Potency factor	YLL, acute tox	Poisoning personyear	Resources (kg)	Biodiversity	Impact cost ($/kg)
Parathion-methyl	000298-00-0	7.14E+01	2.20E–04	2.09E–05	P: 0.118; S: 0.122	2.92E–12	3.43E+01
Paris green	012002-03-8	4.55E+01	1.40E–04	1.33E–05	As: 0.148, Cu: 0.250	1.86E–12	6.28E+02
Pebulate	001114-71-2	8.93E–01	2.75E–06	2.62E–07	S: 0.158	3.66E–14	8.57E+00
Pendimethalin	040487-42-1	9.52E–01	2.93E–06	2.79E–07		3.90E–14	8.56E+00
Pentachlorophenol	000087-86-5	1.25E+01	3.85E–05	3.66E–06		5.12E–13	1.26E+01
Permethrin	052645-53-1	2.00E+00	6.16E–06	5.86E–07		8.19E–14	8.93E+00
Phenthoate	002597-03-7	2.50E+00	7.69E–06	7.33E–07	P: 0.0969, S: 0.200	1.02E–13	9.94E+00
Phenylmercury acetate	000062-38-4	4.17E+01	1.28E–04	1.22E–05	Hg: 0.595	1.71E–12	5.12E+04
Phorate	000298-02-2	5.00E+02	1.54E–03	1.47E–04	P: 0.135, S: 0.417	2.05E–11	1.85E+02
Phosalone	002310-17-0	8.33E+00	2.56E–05	2.44E–06	P: 0.0842; S: 0.174	3.41E–13	1.19E+01
Phosmet	000732-11-6	8.85E+00	2.72E–05	2.59E–06	P: 0.0978, S: 0.202	3.62E–13	1.22E+01
Phosphamidon	013171-21-6	1.43E+02	4.40E–04	4.19E–05	P: 0.103	5.85E–12	5.93E+01
Phoxim	014816-18-3	5.06E–01	1.56E–06	1.48E–07	P: 0.104, S: 0.107	2.07E–14	9.28E+00
Piperophos	024151-93-7	3.09E+00	9.50E–06	9.05E–07	P: 0.0878, S: 0.181	1.26E–13	1.01E+01
Pirimicarb	023103-98-2	6.80E+00	2.09E–05	1.99E–06		2.79E–13	1.06E+01
Pirimiphos-methyl	029232-93-7	6.00E–01	1.85E–06	1.76E–07	P: 0.102, S: 0.105	2.46E–14	9.29E+00
Prallethrin	023031-36-9	2.17E+00	6.69E–06	6.37E–07		8.90E–14	8.99E+00
Prochloraz	067747-09-5	6.25E–01	1.92E–06	1.83E–07		2.56E–14	8.44E+00
Profenofos	041198-08-7	2.79E+00	8.60E–06	8.19E–07	P: 0.0829, S: 0.0856	1.14E–13	9.90E+00
Propachlor	001918-16-7	6.67E–01	2.05E–06	1.95E–07		2.73E–14	8.46E+00

Propanil	000709-98-8	7.14E–01	2.20E–06	2.09E–07		2.92E–14	8.48E+00
Propetamphos	031218-83-4	9.43E+00	2.90E–05	2.77E–06	P: 0.110, S: 0.114	3.86E–13	1.25E+01
Propiconazole	060207-90-1	6.58E–01	2.02E–06	1.93E–07		2.69E–14	8.46E+00
Propoxur	000114-26-1	1.05E+01	3.24E–05	3.09E–06		4.31E–13	1.19E+01
Prosulfocarb	052888-80-9	5.49E–01	1.69E–06	1.61E–07	S: 0.127	2.25E–14	8.44E+00
Prothiofos	034643-46-4	1.08E+00	3.33E–06	3.17E–07	P: 0.0899, S: 0.186	4.43E–14	9.38E+00
Pyraclofos	077458-01-6	4.22E+00	1.30E–05	1.24E–06	P: 0.0859, S: 0.0886	1.73E–13	1.04E+01
Pyrazophos	013457-18-6	2.30E+00	7.08E–06	6.74E–07	P: 0.0831, S: 0.0858	9.41E–14	9.73E+00
Pyrazoxyfen	071561-11–0	6.08E–01	1.87E–06	1.78E–07		2.49E–14	8.44E+00
Pyrethrins	008003-34-7	1.33E+00	4.10E–06	3.91E–07		5.46E–14	8.69E+00
Pyridaben	096489-71-3	1.22E+00	3.75E–06	3.57E–07	S: 0.0877	4.99E–14	8.67E+00
Pyridaphenthion	000119-12–0	1.30E+00	4.00E–06	3.81E–07	P: 0.0912, S: 0.0941	5.32E–14	9.45E+00
Pyroquilon	057369-32-1	3.13E+00	9.62E–06	9.16E–07		1.28E–13	9.32E+00
Quinalphos	013593-03-8	1.61E+01	4.96E–05	4.73E–06	P: 0.104, S: 0.107	6.60E–13	1.48E+01
Quinoclamine	002797-51-5	7.35E–01	2.26E–06	2.16E–07		3.01E–14	8.48E+00
Quizalofop	076578-12-6	5.99E–01	1.84E–06	1.76E–07		2.45E–14	8.44E+00
Quizalofop-p-tefuryl	119738-06-6	9.88E–01	3.04E–06	2.90E–07		4.05E–14	8.57E+00
Rotenone	000083-79-4	1.23E+00	3.77E–06	3.59E–07		5.02E–14	8.66E+00
Simetryn	001014-70-6	5.46E–01	1.68E–06	1.60E–07	S: 0.150	2.24E–14	8.45E+00
Sodium arsenite	007784-46-5	1.00E+02	3.08E–04	2.93E–05	As: 0.326	4.09E–12	1.29E+03
Sodium chlorate	007775–09-9	8.33E–01	2.56E–06	2.44E–07		3.41E–14	8.52E+00

Table 11.2 contd. ...

... Table 11.2 contd.

Substance name	CASRN	Potency factor	YLL, acute tox	Poisoning personyear	Resources (kg)	Biodiversity	Impact cost ($/kg)
Sodium cyanide	000143-33-9	1.67E+02	5.13E−04	4.89E−05		6.82E−12	6.68E+01
Sodium fluoroacetate	000062-74-8	5.00E+03	1.54E−02	1.47E−03	F: 0.190	2.05E−10	1.77E+03
Spiroxamine	118134-30-8	2.00E+00	6.16E−06	5.86E−07		8.19E−14	8.93E+00
Strychnine	000057-24-9	6.25E+01	1.92E−04	1.83E−05		2.56E−12	3.02E+01
Sulfluramid	004151-50-2	1.84E+00	5.67E−06	5.40E−07	S: 0.0607	7.54E−14	8.88E+00
Sulfotep	003689-24-5	2.00E+02	6.16E−04	5.86E−05	P: 0.193, S: 0.199	8.19E−12	8.01E+01
TCA (acid)	000076-03-9	2.50E+00	7.69E−06	7.33E−07		1.02E−13	9.10E+00
Tebuconazole	107534-96-3	5.88E−01	1.81E−06	1.72E−07		2.41E−14	8.43E+00
Tebufenpyrad	119168-77-3	1.68E+00	5.17E−06	4.93E−07		6.88E−14	8.82E+00
Tebupirimfos	096182-53-5	7.69E+02	2.37E−03	2.25E−04	P: 0.0975, S: 0.101	3.15E−11	2.79E+02
Tebuthiuron	034014-18-1	1.55E+00	4.78E−06	4.55E−07	S: 0.140	6.36E−14	8.80E+00
Tefluthrin	079538-32-2	4.55E+01	1.40E−04	1.33E−05	F: 0.317	1.86E−12	2.61E+01
Terbufos	013071-79-9	5.00E+02	1.54E−03	1.47E−04	P: 0.108, S: 0.333	2.05E−11	1.85E+02
Terbumeton	033693-04-8	2.07E+00	6.37E−06	6.07E−07		8.48E−14	8.95E+00
Tetraconazole	112281-77-3	9.70E−01	2.99E−06	2.84E−07	F: 0.204	3.97E−14	9.79E+00
Thallium sulfate	007446-18-6	9.09E+01	2.80E−04	2.66E−05	S: 0.0634, Tl: 0.810	3.72E−12	4.63E+03
Thiacloprid	111988-49-9	2.53E+00	7.77E−06	7.40E−07	S: 0.126	1.03E−13	9.14E+00
Thiobencarb	028249-77-6	7.69E−01	2.37E−06	2.25E−07	S: 0.126	3.15E−14	8.52E+00
Thiocyclam	031895-22-4	3.23E+00	9.93E−06	9.46E−07	S: 0.530	1.32E−13	9.46E+00
Thiodicarb	059669-26-0	1.52E+01	4.66E−05	4.44E−06	S: 0.271	6.20E−13	1.36E+01

Thiofanox	039196-18-4	1.25E+02	3.85E–04	3.66E–05	S: 0.147	5.12E–12	5.22E+01
Thiometon	000640-15-3	8.33E+00	2.56E–05	2.44E–06	P: 0.126, S: 0.390	3.41E–13	1.23E+01
Thiram	000137-26-8	1.79E+00	5.50E–06	5.23E–07	S: 0.533	7.31E–14	8.96E+00
Tralkoxydim	087820-88-0	1.07E+00	3.30E–06	3.14E–07		4.38E–14	8.60E+00
Tralomethrin	066841-25-6	1.18E+01	3.62E–05	3.45E–06		4.82E–13	1.24E+01
Triadimefon	043121-43-3	1.66E+00	5.11E–06	4.87E–07		6.80E–14	8.81E+00
Triadimenol	055219-65-3	1.11E+00	3.42E–06	3.26E–07		4.55E–14	8.62E+00
Triazamate	112143-82-5	1.33E+01	4.10E–05	3.91E–06	S: 0.102	5.46E–13	1.29E+01
Triazophos	024017-47-8	1.22E+01	3.75E–05	3.57E–06	P: 0.099, S: 0.102	4.99E–13	1.33E+01
Trichlorfon	000052-68-6	4.00E+00	1.23E–05	1.17E–06	P: 0.121	1.64E–13	1.06E+01
Triclopyr	055335-06-3	1.41E+00	4.33E–06	4.13E–07		5.77E–14	8.72E+00
Tricyclazole	041814-78-2	3.28E+00	1.01E–05	9.61E–07	S: 0.169	1.34E–13	9.41E+00
Tridemorph	081412-43-3	1.54E+00	4.74E–06	4.51E–07		6.30E–14	8.77E+00
Triflumizole	099387-89-0	1.44E+00	4.43E–06	4.22E–07	F: 0.165	5.89E–14	9.72E+00
Uniconazole	083657-22-1	5.59E–01	1.72E–06	1.64E–07		2.29E–14	8.42E+00
Vamidothion	002275-23-2	9.71E+00	2.99E–05	2.85E–06	P: 0.108, S: 0.223	3.98E–13	1.26E+01
Warfarin	000081-81-2	1.00E+02	3.08E–04	2.93E–05		4.09E–12	4.34E+01
XMC	002655-14-3	1.85E+00	5.68E–06	5.41E–07		7.56E–14	8.87E+00
Xylylcarb	002425-10-7	2.63E+00	8.10E–06	7.71E–07		1.08E–13	9.15E+00
Zeta-cypermethrin	052315-07-8	1.16E+01	3.58E–05	3.41E–06		4.76E–13	1.23E+01
Zinc phosphide	001314-84-7	2.22E+01	6.84E–05	6.51E–06	P: 0.240	9.10E–13	5.30E+01
Ziram	000137-30-4	7.14E–01	2.20E–06	2.09E–07	S: 0.418	2.92E–14	1.84E+01

References

1. Carson, R., *Silent Spring*. 1963, London: Hamish Hamilton.
2. Finizio, A. and S. Villa, Environmental risk assessment for pesticides: A tool for decision making. *Environmental Impact Assessment Review*, 2002. **22**(3): 235–248.
3. Damalas, C.A. and I.G. Eleftherohorinos, Pesticide exposure, safety issues, and risk assessment indicators. *Int. J. Environ. Res. Public Health*, 2011. **8**: 1402–1419.
4. Lamberth, C. et al., *Current Challenges and Trends in the Discovery of Agrochemicals*. Science, 2013. **341**(6147): 742–746.
5. WHO, *The WHO Recommended Classification of Pesticides by Hazard*. 2010: Geneva.
6. WHO, *Global Burden of Disease 2004 Update: Disability Weights for Diseases and Conditions*. 2004, WHO: Geneva.
7. Miller, G.T., in *Sustaining the Earth*. 2004, Thompson Learning, Inc.: Pacific Grove, California. pp. 211–216.
8. Lozano, R. et al., Global and regional mortality from 235 causes of death for 20 age groups in 1990 and 2010: a systematic analysis for the Global Burden of Disease Study 2010. *The Lancet*, 2012. **380**(9859): 2095–2128.
9. Gunnell, D. et al., The global distribution of fatal pesticide self-poisoning: Systematic review. *BMC Public Health*, 2007. **7**(1): 357.
10. Zhang, W., *Global pesticide use: Profile, trend, cost / benefit and more*. Proceedings of the International Academy of Ecology and Environmental Sciences, 2018. **8**: 1–27.
11. Vos, T. et al., Years lived with disability (YLDs) for 1160 sequelae of 289 diseases and injuries 1990–2010: a systematic analysis for the Global Burden of Disease Study 2010. *The Lancet*, 2012. **380**(9859): 2163–2196.

12

Radionuclides

The radionuclides assessed are limited to those considered to have a significant impact from the nuclear energy sector.

12.1 Specification of the Emissions

The emissions are emissions of radioactive substances to air anywhere in the world in 2015 from the nuclear energy sector. The emission unit is TBq.

Impacts from emissions to water are considered of minor importance in the context of this book.

The substances considered are ^{14}C, ^{3}H, ^{129}I, ^{85}Kr, ^{210}PB, ^{210}Po, ^{226}Ra, ^{222}Rn, ^{230}Th, ^{234}U, and ^{238}U.

12.2 Impact Pathways

The nuclides are dispersed in air and people are directly or indirectly exposed to radiation. The collective radiation doses measured in manSv (manSievert) determine the increase of cancer incidence.

12.3 Environmental impact factors

12.3.1 YLL via cancer

The incidence of fatal cancer from nuclear power production is 0.05/manSv [1]. The average loss of life expectancy from all cancers is 14 years [2]. The impact factors in YLL/TBq are shown in Table 12.1.

Table 12.1. Environmental impact factors and values for radioactive nuclides emitted from nuclear energy production.

Nuclide	Collective dose per unit release	YLL/TBq	Cancer personyr/TBq	Impact value, $/TBq
	manSv/TBq			
C-14	85	102	51	1.20E+07
H-3	0.0012	0.00144	0.00072	1.70E+02
I-129	4	4.8	2.4	5.65E+05
Kr-85	2.00E–03	2.40E–03	0.0012	2.83E+02
Pb-210	1	1.2	0.6	1.41E+05
Po-210	1	1.2	0.6	1.41E+05
Ra-226	0.6	0.72	0.36	8.48E+04
Rn-222	0.015	0.018	0.009	2.12E+03
Th-230	30	36	18	4.24E+06
U-234	8	9.6	4.8	1.13E+06
U-238	7	8.4	4.2	9.89E+05

12.3.2 Personyears lived with cancer

The average incidence of radiation-induced cancer is 0.12/manSv [1]. The average length of cancer disability is assumed to be 5 years. The impact factors in personyears/TBq are shown in Table 12.1.

12.4 Summary of Environmental Impact Factors and Values

As shown in table 12.1, the highest impact value comes from C-14. This value is sensitive to the time-period for which the human exposure is assessed. In his case we have used 10000 years, due to the long life-time of C-14.

Edlund [1] also calculated the YLL from normal electricity production in nuclear plants (no accidents). He estimated that 245 YLL was caused per GWa, which is 2.80E–08 YLL/kWh and 2.99E–03 $/kwh.

For comparison, the environmental impact cost from coal-based power plants would be in the order of 0.3 $/kWh, just from CO_2. The exact value depends on energy efficiency, other emissions and whether it is combined heat and power or not.

References

1. Edlund, O., *Estimation of the Years of Lost Life (YOLL) as a consequence of the nuclear fuel cycle*. 2001, Chalmers University of Technology, Swedish Life Cycle Center: Göteborg, Sweden. p. 28. Available from lifecyclecenter.se.
2. Brustugun, O.T., B. Møller and Å. Helland, Years of life lost as a measure of cancer burden on a national level. *Br J Cancer*, 2014. **111**(5): 1014–1020.

13

Littering

Impacts from waste is normally evaluated in terms of the substances that make up the waste. An exception is littering to soil and water, where there is a common impact regardless of the chemical composition.

Littering has long been an environmental issue in terms of littering on land areas [1]. Litter comprise cigarette butts, paper, food wrappers, confection, tissue napkins, miscellaneous plastics, bottles and cans found at public places like road sides and recreational areas. A large portion of individuals contribute to littering.

Lately litter in the seas has been a hot topic. Reports on impacts on marine life are frequent. Pictures of islands of debris are often shown. In a study on marine anthropogenic litter on British beaches [2] it was found that the majority of traceable items originated from land, specifically from public littering.

13.1 Littering on Land

13.1.1 Specification of the emission

Several studies have been made of the type and origin of litter. On number basis, cigarette butts dominate, at least some years ago, when the studies were made, but as smoking is decreasing and cigarette butts are biodegradable, the measure of emission of litter may focus on other issues. There is however no single good measure for litter. Number, surface and weight may be relevant measures. For the impact, number and surface tend to be more relevant than weight, unless the material is toxic, but toxic material is not assessed here. But for product development and waste treatment, weight tends to be more relevant. A correlation between weight and area exists for packaging materials, which are as thin as possible. So, the emission is the mass in kg of thin packaging material or similar thin materials on land anywhere in the world 2015.

13.1.2 Impacts pathways

As litter is removed from many areas, there is an impact on human working capacity. There are examples on injuries from broken glass and cattle that swallow litter, but no quantitative information has been found on these impacts.

13.1.3 Environmental impact factors

13.1.3.1 Decreased working capacity

In 2016 to 2017 UK local authorities paid £682 million to clean up streets from litter [3]. The Clean Europe Network estimates that every year hundreds of thousands of tonnes of litter cover the roads of Europe alone. The costs for cleaning the roads from litter amounts to an estimated €1 billion per year throughout Europe [4]. If "hundreds of thousands" is set to 200 000, then the clean-up costs will be about 5 \$/kg and requiring 1.47 person-hours of working capacity.

The estimated cost and working capacity are for European conditions and for road cleaning. If hand-picked, a piece of litter weighing a few grams may take a few seconds to collect. A kg could then be collected within hours. Besides uncertainty with respect to cost of collection, there is also an uncertainty in the emissions measure with respect to cost of collection. A heavy glass bottle is easier to collect than a small piece of paper. A factor of 3 is assumed as an estimate of the uncertainty.

13.1.4 Summary and monetary valuation of impacts from emissions of litter to land

A best estimate is 5 \$/kg.

13.2 Littering of Sea Waters

According to Nelms et al. [2] most litter on English beaches originates from the land, and most of it is plastics. So, the litter is similar as litter on land, but impacts are different and mitigation methods. Litter in the sea may be transported long distances and interfere with marine life in various ways.

13.2.1 Specification of the emission

The emission is the same as in Sec. 13.1.1, i.e., mass in kg of thin packaging material or similar thin materials emitted to water anywhere in the world 2015.

Primary microplastics is not included in the emissions assessed. Such emissions originate from, e.g., cosmetics and may pass waste water treatment plants. Their effects are largely unknown, although they are found in a large part of present fish catches.

13.2.2 Impacts pathways

Litter reaches sea in three ways: direct, via waterways and via the atmosphere.

Plastic debris in sea have negative effects on wildlife, fishing gears, aquaculture, safety of sea transport, tourism and recreation [5]. Among the environmental goods identified in Chapter 5 (endpoints), human working capacity, fish production capacity, meat production capacity and biodiversity are affected.

Human working capacity (marginal, optional) is affected through work needed on mitigation measures such as cleaning up seashores and repairing equipment.

Fish and meat production capacity may be affected via decreased populations and decreased growth rates. Growth rates may be decreased as plastics are ingested and occupy space for digestion of food. No quantitative estimation of decreased populations or growth rates are found [5].

Some of the species affected by plastic debris and other litter are endangered, like sea turtles. There are however no quantitative estimations on the extent of this impact.

Plastic waste causes $13 billion in annual damage to marine ecosystems, says UN agency (http://www.un.org/apps/news/story.asp?NewsID=48113#.VGDzBMnYfaI).

"It is estimated that 10 to 20 million tons of plastic is finding its way into the world's oceans each year, costing approximately US$ 13 billion per year in environmental damage to marine ecosystems. This includes financial losses incurred by fisheries and tourism as well as time spent cleaning up beaches."

Plastic waste is a big problem for marine environments, but the quantitative knowledge on impacts is low. The cost estimate above does not meet scientific requirements but is an indication of the significance of the problem.

13.2.3 Environmental impact factors

UNEP estimated that in the year 2010, between 4.8 and 12.7 million metric tonnes of plastic found their way into the oceans [5] and the cost for cleaning up shores were estimated to 1500 $/tonne in the APEC region. Adjusted to OECD conditions using average GDP/capita ratio, the cost will be 42000/14800 = 2.83 times higher, i.e., 4250 $/ton or 4.25 $/kg. This corresponds to an impact on working capacity of 4.25/107067 = 3.97E–05 personyears/kg litter.

The estimated cost and working capacity are for APEC conditions and for seashore cleaning. GDP varies very much in the APEC countries, and is not clear where the costs were measured. Besides uncertainty with respect to cost of collection, there is also an uncertainty in the emissions measure with respect to cost of collection. A heavy glass bottle is easier to collect than a small piece of paper.

There is a large uncertainty about the value of impacts on production capacity of fish and meat and on biodiversity. Even if costs of cleaning may be seen as a valuation of every negative impact of the litter, this valuation is very uncertain.

A factor of 5 is assumed as an estimate of the uncertainty.

To conclude: the best estimate of the impact cost of litter emitted to the sea is 4.25 $/kg with an uncertainty of a factor of 5.

References

1. Schultz, P.W. et al., Littering in context: personal and environmental predictors of littering behavior. *Environment and Behavior*, 2013. **45**(1): 35–59.
2. Nelms, S.E. et al., Marine anthropogenic litter on British beaches: A 10-year nationwide assessment using citizen science data. *Science of The Total Environment*, 2017. **579**: 1399–1409.
3. UK Department for Environment, F.a.R.A. *Litter and littering in England 2016 to 2017*. 2018 [cited 2019 May 8]; Available from: https://www.gov.uk/government/publications/litter-and-littering-in-england-2016-to-2017/litter-and-littering-in-england-2016-to-2017.
4. Network, C.E. *The Litter Challenge, Facts and Costs*. 2019 [cited 2019 May 8]; Available from: https://www.cleaneuropenetwork.eu/en/facts-and-costs/aup/.
5. Kershaw, P.J., *Marine plastic debris and microplastics – Global lessons and research to inspire action and guide policy change*. 2016, United Nations Environment Programme: Nairobi.

14
Land Use

Land use as an environmental impact category is different from emissions and resource extractions. It is not a flow. It is an activity, which encompasses several flows and physical impacts on the environment. However, it has proven practical to handle is as a flow with the dimension area*time, because detailed data on all physical and chemical interactions are seldom available on a general level, as in product and process development. When addressing environmental issues in local land use contexts, the picture is different. Then, the monetary valuations for land use categories developed here cannot be used, although some of the analytical structure can be used.

In LCA, land use is often categorized in two groups, land use and land transformation. Here, land transformation is seen as part of land use, and its impacts are allocated to and distributed over the land use temporal extension.

Here, the land use categorization used by IUCN will be used. In turn, it builds on UN categorization. The categories used are summarized in Table 14.1.

14.1 Residential and Commercial Developments in Cities with > 0.5 Million Inhabitants

14.1.1 Specification of the land use category

Residential and commercial developments in cities with > 0.5 million inhabitants involve hard-making of surfaces, drainage systems, high buildings, parks and traffic systems. Variations are significant between

Table 14.1. Land use categories use by IUCN. A further categorization is made between large and small cities because of different degrees of heat island formations.

Land use activity	Unit
Residential & commercial developments in cities > 0.5 million inhabitants	
Housing and urban areas, arable land	m^2 year
Housing and urban areas, forestland	m^2 year
Housing and urban areas, impediment	m^2 year
Commercial & industrial areas, arable land	m^2 year
Commercial & industrial areas, forestland	m^2 year
Commercial & industrial areas, impediment	m^2 year
Tourism & recreational areas	m^2 year
Residential & commercial developments in rural and cities < 0.5 million inhabitants	
Housing and urban areas, arable land	m^2 year
Housing and urban areas, forestland	m^2 year
Housing and urban areas, impediment	m^2 year
Commercial & industrial areas, arable land	m^2 year
Commercial & industrial areas, forestland	m^2 year
Commercial & industrial areas, impediment	m^2 year
Tourism & recreational areas	m^2 year
Agriculture and Aquaculture	
Annual&perennial non-timber crops	m^2 year
Wood & pulp plantations	m^2 year
Livestock farming and ranching	m^2 year
Marine and freshwater aquaculture	kg produced
Energy production and mining	
Oil and gas drilling	kg produced
Mining and quarrying	m^2 year
Renewable energy	m^2 year
Transportation and service corridors	
Roads and railroads	m^2 year
Utility and service lines	m^2 year
Biological resource use	
Logging and wood harvesting	m^2 year

cities around the world. The land use category valued here is an average urban complex with more than 500 000 inhabitants anywhere in the world in 2015.

14.1.2 Impacts pathways

Impacts on environmental goods and services are caused by two mechanisms: obstruction of natural ecosystems and development of urban heat islands (Table 14.2).

Table 14.2. Environmental goods affected by residential and commercial developments in cities with > 0.5 million inhabitants.

Environmental good	Impact indicator	Unit	Pathway
human health	YLL	personyears	urban heat islands
human health	working capacity	personhours	heat stress
crop	production capacity	kg	land occupation
wood	production capacity	m^3	land occupation
drinking water	production capacity	m^3	land occupation
biodiversity	share of threat to redlisted species	dimensionless	land occupation

14.1.3 Environmental impact factors

14.1.3.1 YLL via heat stress and cold moderation

It has long been known that an increased temperature during heat waves causes excess mortality in urban areas. The effect becomes statistically significant at daily average values around 30°C [1]. Persons older than 65 years suffering from cardiovascular diseases are at high risk.

In Sec. 6.1.3.1 the YLL via heat stress and cold moderation was determined from population and temperature statistics. The YLL was determined for a 1.3 degrees temperature increase. In large urban areas there is an extra 1.5 degrees temperature increase due to heat islands [2]. If using the model in Sec. 6.1.3.1 the extent of extra YLL will be 4.51E+05 YLL per year for the 25% of the global population living in cities with more than 500 000 inhabitants. The areas of cities with more than 500000 inhabitants is estimated to be 4.09E+05 km^2 from the average population density and number of people living in these cities. The environmental impact factor thus becomes 4.51E+05/4.09E+11 = 1.10E–06 YLL/m^2 year.

There is a large variation in the environmental impact factor depending on latitude. In high latitudes the impact factor may be negative. Considering the variations between YLL from heat stress and cold moderations, the uncertainty is estimated as ± 6.5E–07 YLL/m^2 year, as a standard deviation in a normal distribution.

14.1.3.2 Decreased working capacity from heat stress

Population density in urban areas varies between 1800 and 44000/km^2 (*New Geography*, February 4, 2015). The average population density is 4400/km^2 [3] resulting in an extra 1.5 degrees on top of RCP6, for urban heat islands [2] resulting in an RCP 8.5 scenario, which will decrease working capacity with an extra 5% (estimated from Dunnes et al. [4]). As about 25% of the global population live in cities with more than 500000 inhabitants, the global average will be 1.25% decrease. The impact per m^2 is assumed to be the same as the contribution per m^2, i.e., the damage caused is the same as the damage experienced per m^2.

The environmental impact factor therefore is 4400/1000000*0.0125*365*16 = 3.21E–01 personhours/m^2 year.

There is a large variation in the environmental impact factor depending on latitude. In high latitudes the impact factor may be negative. Considering the variations between YLL from heat stress and cold moderations, the uncertainty is estimated to ± 1.9E–01 personyears/m^2 year, as a standard deviation in a normal distribution.

14.1.3.3 Decreased crop production capacity from land occupation

In case the city is built on agricultural land the crop production capacity is decreased. Assuming an average crop yearly production capacity of 6 ton/hectare or 0.6 kg/m^2 year, the environmental impact factor is the loss of that productivity per year, i.e., 0.6 kg/m^2 year.

An uncertainty of a factor of 2 is assumed, considering variations in productivity in different areas in the world.

14.1.3.4 Decreased wood production capacity from land occupation

In case the city is built on forest land the wood production capacity is decreased. Assuming an average yearly wood production capacity of 6 m^3/hectare or 0.0006 m^3/m^2 year, the environmental impact factor is the loss of that productivity per year, i.e., 6E–04 m^3 wood/m^2 year.

An uncertainty of a factor of 2 is assumed, considering variations in productivity in different areas in the world.

14.1.3.5 Decreased drinking water production capacity from land occupation

The quality of run-off water from precipitation in cities is normally too low to allow the use of water for drinking purposes.

The average production of drinking water in the world is estimated to be 0.308 m^3/m^2 per year, which is average precipitation minus evaporation.

Total global runoff (excl. Antarctica) is about 40000 km^3/year and total land area (ex. Antarctica) is 130 million km^2. 40000/130000000 = 3.08E–04 km, which is 0.308 m or 0.308 m^3/m^2.

An uncertainty of a factor of 3 is assumed, considering variations in precipitation in different areas in the world.

14.1.3.6 Decreased biodiversity from land occupation

Total urban area globally is 400000 km^2 (FAO).

The total area of commercial and industrial activities is estimated to be 200000 km^2.

The share of red-listed species threatened by residential and commercial developments is 0.104 and 0.026 from commercial and industrial activities [5].

The environmental impact factor is then 0.104/4.0E+11 = 2.60E–13 shares of threat to red-listed species/m^2 year from residential and commercial activities and 0.026/2.0E+11 = 1.3E–13 shares of threat to red-listed species/m^2 year from commercial and industrial activities.

The uncertainty is estimated to a factor of 3 depending on varying amounts of green areas.

14.1.4 Summary of environmental impact factors and monetary valuation of residential & commercial developments in cities with > 0.5 million inhabitants

The impact factors and monetary values vary slightly depending on the type of residential and commercial development area. In Table 14.3 the data are given from housing and urban areas. Table 14.4 show aggregated data for all types of areas in cities with more than half a million inhabitants.

14.2 Residential & Commercial Developments in Cities with < 0.5 Million Inhabitants

In cities with 100000 inhabitants, heat island temperature differences are typically 2/3 of the ones in a 1 million city, resulting in about 1 degrees temperature increase [6] and 2/3 of the environmental impact factors as linear impact models are used (Table 14.5).

Table 14.3. Environmental impact factors and monetary impact values for housing and urban areas built on arable land. The total impact value corresponds to that of the first row in Table 14.4.

Environmental good	Impact indicator	Unit	Pathway	Environmental impact factor	Uncertainty	Indicator value ($/unit)	Impact value ($/m^2 year)
human health	YLL	personyears	urban heat islands	1.10E–06	± 6.5E–07	107,067	1.18E–01
human health	working capacity	personhours	heat stress	3.21E–01	± 1.9E–01	30	9.64E+00
crop	production capacity	kg	land occupation	6.00E–01	2	0.289	1.73E–01
wood	production capacity	m^3	land occupation	0	2	72	0
drinking water	production capacity	m^3	land occupation	3.08E–01	3	1.87	5.76E–01
biodiversity	share of threat to redlisted species	dimensionless	land occupation	2.60E–13	3	7.61E+10	1.98E–02
						sum	1.05E–01

Table 14.4. Environmental impact factors and monetary impact values for residential & commercial developments in cities with > 0.5 million inhabitants.

Land use activity	**Unit**	**Environmental impact factors**						
		YLL/unit	**Working capacity (p-yr/unit)**	**Crop (kg/unit)**	**Wood (m3/unit)**	**Drinking water (m3)**	**Biodiversity (dimension-less)**	**Impact value ($/m² year)**
Residential & commercial developments in cities > 0.5 million inhabitants								
Housing and urban areas, arable land	m² year	1.10E–06	3.21E–01	6.00E–01		3.08E–01	2.60E–13	1.05E+01
Housing and urban areas, forestland	m² year	1.10E–06	3.21E–01		6.00E–04	3.08E–01	2.60E–13	1.04E+01
Housing and urban areas, impediment	m² year	1.10E–06	3.21E–01			3.08E–01	2.60E–13	1.03E+01
Commercial & industrial areas, arable land	m² year	1.10E–06	3.21E–01	6.00E–01		3.08E–01	1.30E–13	1.05E+01
Commercial & industrial areas, forestland	m² year	1.10E–06	3.21E–01		6.00E–04	3.08E–01	1.30E–13	1.04E+01
Commercial & industrial areas, impediment	m² year	1.10E–06	3.21E–01			3.08E–01	1.30E–13	1.03E+01
Tourism & recreational areas	m² year	1.10E–06					9.25E–14	1.25E–01

Table 14.5. Environmental impact factors and monetary impact values for residential & commercial developments in cities > 0.5 million inhabitants.

Land use activity	Unit	Environmental impact factors						
		YLL/unit	Working capacity (p-yr/unit)	Crop (kg/unit)	Wood (m3/unit)	Drinking water (m3)	Biodiversity (dimension-less)	Impact value ($/m^2 year)
Residential & commercial developments in rural and cities < 0.5 million inhabitants								
Housing and urban areas, arable land	m^2 year	7.35E–07	2.14E–01	6.00E–01		3.08E–01	2.60E–13	7.27E+00
Housing and urban areas, forestland	m^2 year	7.35E–07	2.14E–01		6.00E–04	3.08E–01	2.60E–13	7.14E+00
Housing and urban areas, impediment	m^2 year	7.35E–07	2.14E–01			3.08E–01	2.60E–13	7.10E+00
Commercial & industrial areas, arable land	m^2 year	7.35E–07	2.14E–01	6.00E–01		3.08E–01	1.30E–13	7.26E+00
Commercial & industrial areas, forestland	m^2 year	7.35E–07	2.14E–01		6.00E–04	3.08E–01	1.30E–13	7.13E+00
Commercial & industrial areas, impediment	m^2 year	7.35E–07	2.14E–01			3.08E–01	1.30E–13	7.09E+00
Tourism & recreational areas	m^2 year	7.35E–07					9.25E–14	8.57E–02

Table 14.6. Environmental impact factors and monetary impact values for various land use categories.

Land use activity	Unit	Environmental impact factors						
		YLL/unit	Working capacity (p-yr/unit)	Crop (kg/unit)	Wood (m^3/unit)	Drinking water (m^3)	Biodiversity (dimension-less)	Impact value (\$/m^2 year)
Agriculture and Aquaculture								
Annual & perennial non-timber crops	m^2 year						1.07E–14	8.16E–04
Wood & pulp plantations	m^2 year						2.00E–14	1.52E–03
Livestock farming and ranching	m^2 year						3.33E–15	2.54E–04
Marine and freshwater aquaculture	kg produced						6.27E–14	4.77E–03
Energy production and mining								
Oil and gas drilling	kg produced						4.99E–16	3.79E–05
Mining and quarrying	m^2 year				6.00E–04	3.08E–01	6.86E–14	6.24E–01
Renewable energy	m^2 year						2.95E–15	2.24E–04
Transportation and service corridors								
Roads and railroads	m^2 year			6.00E–01		3.08E–01	4.01E–12	1.05E+00
Utility and service lines	m^2 year				6.00E–04		1.71E–15	4.33E–02
Biological resource use								
Logging and wood harvesting	m^2 year						3.76E–15	2.86E–04

14.3 Other Land Use

Based on IUCN's data on threats to endangered species and area estimations environmental impact factors and environmental impact values have been calculated for several other land use categories (Table 14.6).

References

1. Weihe, Urban Climatology and its Applications with Special Regard to Tropical Areas, in *Proceedings of the technical conference, Mexico D.F, 26–30 November 1984. 1986, WMO: Geneva.*
2. Peng, S. et al., *Response to Comment on "Surface Urban Heat Island Across 419 Global Big Cities". Environmental Science & Technology, 2012.* **46**(12): 6889–6890.
3. *Demographia World Urban Areas: Built-Up Urban Areas or World Agglomerations. 2015, Wendell Cox Consultancy: http://www.demographia.com/db-worldua.pdf.*
4. Dunne, J.P., R.J. Stouffer and J.G. John, *Reductions in labour capacity from heat stress under climate warming. Nature and Climate Change, 2013.* **3**: 4.
5. *IUCN Redlist. 2014.*
6. *Oke,* Urban Climatology and its Applications with Special Regard to Tropical Areas., *in Proceedings of the technical conference, Mexico D.F., 26–30 November 1984* . 1986, WMO: Geneva.

15

Global Impacts

The environmental impact values estimated in the preceding chapters are difficult to validate, except through the methodological process by which they were determined. An additional test on their validity is to compare the added monetary values of all emissions and resource extractions in the world with the global GDP. Is it of reasonable size?

If sustainability is the capability for long-term welfare, the prospect to manage the natural capital depends on the relation between economic performance and environmental performance. Sufficient environmental resources are critical for human welfare. If the environment is impoverished, more traditional economic resources are needed to satisfy human basic needs.

In Table 15.1 the monetary values determined in Chapters 6–11 are multiplied with the global emissions and resource extractions.

The total decrease of the natural capital 2015 was in this way determined to 8.92E+13 $. The global GDP 2015 was 7.56E+13 US$. It looks as we are destroying more values than we are creating. However, in valuing clean air and water via impacts on human health, values for working capability of OECD inhabitants were used. This means that the environmental impact values are not directly comparable with the global GDP. A better comparison for the natural capital value would be the working capacity of an average OECD inhabitant multiplied with the global population: 107067*7.2E+09 = 7.71E+14 $. This is 10.2 times higher than the global GDP. When applying 7.71E+14 $ for values created globally, the decrease of natural capital is now 11.6% of this new economic capability. But there is still a problem in that we are borrowing 11.6% each year from the natural capital, while we are not increasing our economic capability more than a few %. This is not sustainable.

A safe sustainable development would mean that the natural capital increased or at least did not decrease. A less safe sustainable development would mean that the decrease of the natural capital at least was less than the economic growth.

The ten flows with highest monetary values are shown in Table 15.2.

- Not surprisingly, CO_2 is the most expensive emission in terms of its decrease of natural capital. Number 2 is PM2.5. Nor this is surprising. PM2.5 is regarded by WHO as one of the most severe health issues, even worse than smoking.
- Number 3 is depletion of gold resources. This will probably be questioned as the valuation only includes natural Au concentrates, not the gold in the technical sphere. But even if the recycling rates are high, losses on a percent level will ultimately deplete those resources as well.
- Number 4 is Rhodium. Rhodium is so rare in earth's crust, that one may question if it ever will be mined from such very dilute states. But even so, it is a loss of a resource, that may have qualities we are not yet aware of.
- Number 5 and 7 is land use in urban areas. The reason for the high environmental impact values is the contribution to urban heat islands. Adding a few extra degrees on top of global warming causes severe health effects including loss of working capacity.
- Number 6 is Antimony, Sb. In Life Cycle Impact Assessment, Sb is often used as a reference for the impact category "resource depletion". Resource flows are quantified as "Sb-equivalents".
- Number 8 is methane, CH_4, which is the second most important climate gas after CO_2.
- Number 9 is the most used metal: iron, Fe. Even if much of iron is recycled, mining is still extensive.
- Oil is number ten on the list. There is a well-known concern about its depletion, and the forthcoming economic consequences.

Values in Table 15.1 and 15.2 represents "business as usual"—scenarios. Mitigation and adaption may decrease the impact costs. It is striking though, that all are global or regional issues in that they cannot be handled on a local scale. This is in line with the historical development of environmental care. Local issues are handled first. Toxicity issues are addressed more easily than resource issues [9].

Table 15.1. Global decrease of natural capital from emissions and resource extractions. Data marked a is from IPCC AR5 WG1, b from [1], c from UN, d from FAO [2], e from EUROSTAT, f from World Bank, g from Galloway et al. [3], h from Liu et al. [4], i from Brown et al. [5], j from US Energy Information Administration, k from USGS, l from Moss et al. [6], m from http://www.mmta.co.uk/metals/Er, n from http://www.lenntech.com/periodic/elements/ho.htm, o from http://metals.about.com/od/properties/a/Metal-Profile-Iridium.htm, p from Wikipedia, q from International Manganese Institute, http://www.manganese.org/about-mn/production, r from www.encyclopedia.com/topic/osmium.aspx, s from www.iamgold.com/files/REE101_April_2012.pdf, t from http://www.statista.com/statistics/273645/global-mine-production-of-platinum/, u from Emsley [7], v from the same book as u, but from pages 371–374 and x from World Nuclear Association [8].

Flow	Unit	Value/unit	Global flow	Reference	Value, $
CO_2	kg	3.56E–01	3.26E+13	a	1.16E+13
CH_4	kg	1.01E+01	3.33E+11	a	3.37E+12
N_2O	kg	9.46E+01	1.00E+10	a	9.46E+11
NOx	kg	-3.28E+01	1.23E+11	a	–4.04E+12
SO_2	kg	-1.04E+01	1.30E+11	a	–1.35E+12
VOC	kg	1.01E+01	1.27E+11	a	1.28E+12
PM2.5	kg	2.68E+02	3.80E+10	b	1.02E+13
Urban land use > 0.5 million	m^2 yr	1.05E+01	4.68E+11	c	4.91E+12
Urban land use < 0.5 million	m^2 yr	7.20E+00	4.68E+11	c	3.37E+12
Agriculture, cropland	m^2 yr	8.16E–04	1.38E+13	d	1.13E+10
Agriculture, pastures	m^2 yr	2.54E–04	3.35E+13	d	8.51E+09
Forestry, plantations	m^2 yr	1.52E–03	2.50E+12	d	3.80E+09
Logging and wood harvesting	m^2 yr	2.86E–04	4.00E+13	d	1.14E+10
Oil and gas drilling	kg	3.79E–05	3.53E+12		1.34E+08
Mining and quarrying	m^2 yr	6.24E–01	5.51E+11	e	3.44E+11
BOD to freshwater	kg	3.35E–04	1.60E+10	f	5.36E+06
Ntot to freshwater	kg	2.40E–03	1.18E+11	g	2.83E+08
Ptot to freshwater	kg	4.55E–02	3.80E+10	h	1.73E+09
Oil	kg	8.00E–01	4.01E+12	i	3.21E+12

Table 15.1 contd. ...

... Table 15.1 contd.

Flow	Unit	Value/unit	Global flow	Reference	Value, $
Coal	kg	2.84E–01	7.26E+12	j	2.06E+12
Natural gas	kg	5.00E–01	2.43E+12	i	1.21E+12
Ag	kg	1.15E+05	2.00E+07	k	2.30E+12
Al	kg	1.75E–01	4.70E+10	i	8.23E+09
As	kg	3.84E+03	4.57E+07		1.75E+11
Au	kg	3.20E+06	2.60E+06	i	8.32E+12
B	kg	1.00E+01	4.75E+09	k	4.75E+10
Bi	kg	4.43E+04	8.90E+06	k	3.94E+11
Cd	kg	5.88E+04	2.20E+07	l	1.29E+12
Ce	kg	9.00E+01	2.40E+07	k	2.16E+09
Cr	kg	4.34E+01	5.25E+09		2.28E+11
Co	kg	2.26E+02	1.30E+08		2.94E+10
Cs	kg	7.80E+02	3.00E+04	k	2.34E+07
Cu	kg	1.44E+02	1.80E+10	l	2.59E+12
Dy	kg	1.65E+03	1.20E+06		1.97E+09
Er	kg	2.50E+03	9.40E+05	m	2.35E+09
Eu	kg	6.55E+03	2.70E+05	m	1.77E+09
F	kg	6.00E+00	2.20E+09	k	1.32E+10
Fe	kg	1.10E+00	3.00E+12	l	3.30E+12
Ga	kg	3.39E+02	1.60E+05	l	5.42E+07
Gd	kg	1.52E+03	7.50E+06	m	1.14E+10
Ge	kg	3.60E+03	1.21E+05		4.36E+08
Hf	kg	9.93E+02	8.00E+04	l	7.94E+07
Hg	kg	8.60E+04	1.90E+06		1.63E+11
Ho	kg	7.20E+03	1.00E+04	n	7.20E+07
In	kg	1.15E+05	1.35E+06	l	1.56E+11
Ir	kg	2.62E+08	4.00E+03	o	1.05E+12
La	kg	1.92E+02	1.25E+07	m	2.40E+09
Li	kg	1.00E+01	2.78E+07	i	2.78E+08
Lu	kg	1.80E+04	9.00E+03	p	1.62E+08
Mn	kg	5.58E+00	1.77E+10	q	9.88E+10
Mo	kg	3.84E+03	2.34E+08	l	8.99E+11
Nb	kg	4.80E+02	5.10E+07	k	2.45E+10
Nd	kg	2.22E+02	7.00E+06	p	1.55E+09

Table 15.1 contd. ...

Table 15.1 contd. ...

Flow	Unit	Value/unit	Global flow	Reference	Value, $
Ni	kg	1.36E+02	1.90E+09	i	2.59E+11
Os	kg	1.15E+08	5.00E+02	r	5.76E+10
P	kg	8.20E+00	4.48E+10	k	3.67E+11
Pb	kg	3.39E+02	5.30E+09	i	1.80E+12
Pd	kg	1.09E+07	2.22E+05	p	2.41E+12
Pr	kg	8.11E+02	1.13E+07	s	9.13E+09
Pt	kg	9.60E+06	1.90E+05	t	1.82E+12
Rb	kg	5.24E+01	2.00E+03	k	1.05E+05
Re	kg	1.44E+07	3.67E+04	k	5.28E+11
Rh	kg	3.20E+08	2.50E+04		8.00E+12
Ru	kg	1.92E+08	1.20E+04	u	2.30E+12
S	kg	2.00E–01	6.80E+10	k	1.36E+10
Sb	kg	2.88E+04	1.63E+08	k	4.69E+12
Sc	kg	4.11E+02	1.30E+03	p	5.36E+05
Se	kg	5.00E+01	3.25E+06	l	1.63E+08
Sm	kg	1.28E+03	7.00E+05	v	8.96E+08
Sn	kg	6.98E+02	2.61E+08	l	1.82E+11
Sr	kg	2.00E–01	3.80E+08	k	7.60E+07
Ta	kg	5.76E+03	5.90E+05	k	3.40E+09
Tb	kg	9.00E+03	4.50E+05	s	4.05E+09
Te	kg	5.76E+06	5.00E+05	l	2.88E+12
Ti	kg	1.40E+00	1.80E+07		2.53E+07
Tl	kg	5.76E+03	1.00E+04	k	5.76E+07
Tm	kg	1.75E+04	5.00E+04	u	8.73E+08
U	kg	5.38E+02	5.94E+04	x	3.20E+07
V	kg	5.38E+01	5.60E+07		3.01E+09
W	kg	5.76E+03	7.50E+07	m	4.32E+11
Y	kg	2.62E+02	5.59E+06	k	1.46E+09
Yb	kg	2.62E+03	5.00E+04	p	1.31E+08
Zn	kg	4.61E+01	1.35E+10	k	6.22E+11
Zr	kg	3.03E+01	1.62E+09	k	4.91E+10
			sum		8.47E+13

Table 15.2: Ten flows with highest monetary values.

Flow	Unit	Value/unit	Global flow	Value, $
CO_2	kg	3.56E–01	3.26E+13	1.16E+13
PM2.5	kg	2.68E+02	3.80E+10	1.02E+13
Au	kg	3.20E+06	2.60E+06	8.32E+12
Rh	kg	3.20E+08	2.50E+04	8.00E+12
Urban land use > 0.5 million	m^2 yr	1.05E+01	4.68E+11	4.91E+12
Sb	kg	2.88E+04	1.63E+08	4.69E+12
CH4	kg	1.01E+01	3.33E+11	3.37E+12
Urban land use < 0.5 million	m^2 yr	7.20E+00	4.68E+11	3.37E+12
Fe	kg	1.10E+00	3.00E+12	3.30E+12
Oil	kg	8.00E-01	4.01E+12	3.21E+12

References

1. *EDGAR, Emissions Database for Global Athmospheric Research.* 2017, EU JRC.
2. *FAOSTAT* 2019, FAO.
3. Galloway, J.N. et al., Nitrogen cycles: Past, present, and future. *Biogeochemistry*, 2004. **70**(2): 153–226.
4. Liu, Y. et al., Global phosphorus flows and environmental impacts from a consumption perspective. *Journal of Industrial Ecology*, 2008. **12**(2): 229–247.
5. Brown, T.J. et al., *World Mineral Production 2008-2012*, ed. B.G. Survey. 2014, Nottingham: Keyworth.
6. Moss, R.L. et al., The potential risks from metals bottlenecks to the deployment of Strategic Energy Technologies. *Energy Policy*, 2013. **55**: 556–564.
7. Emsley, J., Ruthenium, In: *Nature's Building Blocks: An A-Z Guide to the Elements*. 2003, Oxford University Press: Oxford. pp. 368–370.
8. Association, W.N. *World Uranium Mining Production 2013*. 2015 [cited 2015 Feb 12]; Available from: http://www.world-nuclear.org/info/Nuclear-Fuel-Cycle/Mining-of-Uranium/World-Uranium-Mining-Production/.
9. Harremoës, P. et al., *Late lessons from early warnings: The precautionary principle 1896–2000*. 2002, European Environmental Agency.

16

Applications in Product and Process Development

The monetary values for emissions and resource use that were estimated in previous chapters were primarily intended for product and process development. Products and processes are here to be understood in a wide meaning. It could be a window cleaning service, or it could be a new production plant. To illustrate the use of monetary values of environmental impacts in product and process development, it works well with a simple example: a small container, made either of steel or of polycarbonate (PC). The example is analyzed in two ways. (1) in the context of LCA or CBA, where emissions and resources are quantified. (2) In parallel with economic evaluation, where costs for materials and processes are aggregated.

16.1 Valuation of LCA or CBA Data

In order to keep focus on valuation issues, technical issues like design, transports and waste handling processes are not discussed here. We just note the emissions and resource requirements (Table 16.1). The example is not a general assessment on environmental performance of materials.

The cost of producing the container is the same for both alternatives, 0.08 \$. Steel is cheaper than PC but the PC alternative weighs less.

The environmental impact values are similar for the two alternatives, with a slightly higher value for the PC case. The impact values are about half of the price/consumer value. The "real price" is thus around 1.2 \$ for the container, but some of it is paid by people somewhere else and at some other time. If applying the polluter pays principle, governments should add

Table 16.1. Example of environmental impact costs from alternative container designs.

	Impact value ($/kg)	Steel (kg emission or resource)	PC (kg emission or resource)	Steel impact value ($)	PC impact value ($)
CO_2	3.56E–01	3.90E–02	6.11E–02	1.39E–02	2.18E–02
NOx	–3.28E+01	2.00E–05	1.20E–04	–6.56E–04	–3.94E–03
SO_2	–1.04E+01	–1.40E–05	1.50E–04	1.46E–04	–1.56E–03
PM2.5	2.68E+02	2.00E–05	2.00E–05	5.36E–03	5.36E–03
CH4	1.01E+01	1.00E–04	6.80E–04	1.01E–03	6.89E–03
NMVOC	1.01E+01	5.00E–05	4.00E–05	5.05E–04	4.04E–04
Oil	8.00E–01	1.50E–03	5.40E–03	1.20E–03	4.32E–03
Coal	2.84E–01	4.40E–03	6.50E–03	1.25E–03	1.85E–03
Natural gas	5.00E–01	9.31E–03	1.25E–02	4.65E–03	6.25E–03
Fe	1.10E+00	7.00E–03	0.00E+00	7.70E–03	0.00E+00
			sum	3.51E–02	4.13E–02

a tax or take some other measure that make the price be 1.2 $ instead of 0.8. If not done, suppliers of the containers can regard 0.4 $ as an economic risk. If the cost can be transferred to the customer, the risk will be small to the supplier, but if not, their profit may turn to a loss.

In both cases, the impact cost estimates motivate suppliers to take measures to decrease these costs. Not only will their economic risk decrease, their competitive strength will increase. More and more purchasers, in particular public purchasers, demand "green" products and services.

For the steel container Table 16.1 indicates that focus should be on decreased emissions of CO_2 and use of fossil and iron mineral resources to decrease the impact cost. For the polycarbonate container, focus should be on decreased emissions of CO_2 and use of fossil mineral resources to decrease the impact cost. The PM2.5 environmental monetary impact values may be different for the two alternatives, so even if the emissions are the same, the values can be different.

The uncertainty in estimating monetary environmental impact values may alter the ranking of the alternatives. Most of the emissions and resources used for the containers have the same values for both alternatives. So even if they are uncertain, errors do not change the ranking between the alternatives with respect to a specific emission, but it may do so for the added values. To find out how certain the ranking between the alternatives are a Monte Carlo simulation can be made. Adding random errors to each emission and resource value and create new values that are log-normal distributed with the standard deviation describing the uncertainty, several new results are obtained for the total environmental damage cost of the containers. Such a simulation with 200 calculations, are shown in Figure 16.1. The value for the steel container minus the value for the PC container is shown on the Y-axis. This means that the steel container is less impacting on the environment in 75% of the calculations and the PC container is preferable in 25% of the cases.

A correct answer to the question of which of the two alternatives that has the lowest environmental impact cost is thus not "the steel container" but "the steel container with 75% probability".

This means that if you invest 100 $ for improving the environment and the probability is 25% that you chose the wrong alternative, the average value of your investment is only about 50 $. Uncertainty has a price. In this case it is 50 $.

To improve the precision in the analysis, the data contributing most to the uncertainty in ranking can be found by the indicator E/σ, where E is the change you can do in a value before the ranking of alternatives changes, and σ is the uncertainty. The larger the E/σ, the more it contributes to the uncertainty.

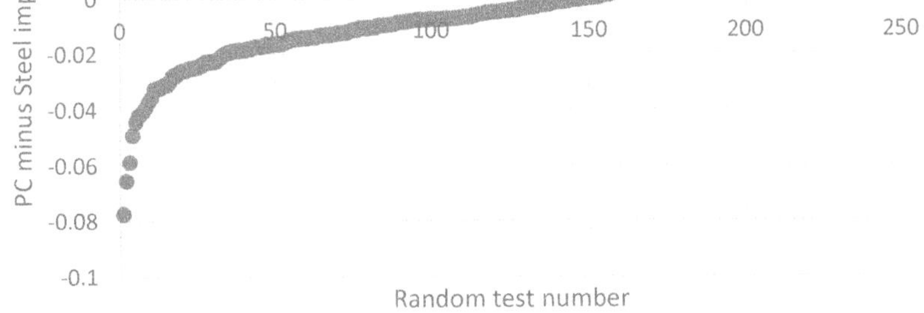

Figure 16.1. Results from Monte Carlo simulation.

16.2 In Parallel with Conventional Cost Assessment

In developing a container, its production costs are central for the success of the product. The cost can be the sum of material costs, processing and distribution cost. In a societal perspective the life cycle cost is of interest. Then, there are also costs of use and end-of-life processing. Conventional as well as external, environmental costs are of interest. Using monetary values for environmental impacts in production, use and end-of-life handling in parallel with conventional cost assessments is much less time-consuming and expertise-demanding than LCA and CBA. But then, a list of environmental prices is needed. In Table 16.2 such a price list is given for the materials and processes used in the containers' life cycles.

The price list in Table 16.2 may be used in a simple spreadsheet by the product developer as in Table 16.3.

Table 16.2. Price list for materials and processes used in containers' life cycles.

Process	Material	Unit	Impact value ($/unit)
manufacturing	steelsheet	kg	2.13
manufacturing	polycarbonate	kg	4.26
stamping	steel	kg	0.05
injection moulding	polycarbonate	kg	0.06
high pressure washing	container	occasion	6.00E–05
landfill	steelsheet	kg	6.00E–07
landfill	polycarbonate	kg	2.00E–06
material reuse	steel	kg	–2.00E+00
material reuse	polycarbonate	kg	–2.13E+00
waste incineration	steel	kg	0
waste incineration	polycarbonate	kg	7.40E–01

Table 16.3. Simplified environmental impact cost calculation for product life cycles.

Process		Production			Use			Reuse-Material			Incineration			Landfill			
		Unit cost	Amount	Impact cost	Unit cost	Amount	Impact cost	Unit cost	Amount	Impact cost	Unit cost	Amount	Impact cost	Unit cost	Amount	Impact cost	Impact cost
	Unit	($)		($)	($)		($)	($)		($)	($)		($)	($)		($)	($)
Manu-facturing of steelsheet	kg	2.13	0.1	0.213			0			0			0			0	0.213
Stamping	kg	0.05	0.1	0.005			0			0			0			0	0.005
High pressure washing	occation			0.006	6.00E–05	100	6.00E–03			0			0			0	0.012
Waste handling	kg			0			0	–2.00	0.095	-0.19	0	0.005	0	2.00E–06	0.005	1E-08	–0.19
				0			0			0			0			0	0
				0			0			0			0			0	0
				0			0			0			0			0	0
				0			0			0			0			0	0
				0			0			0			0			0	0
SUM				0.224			0.006			-0.19			0			1E-08	0.04

17

Reflections and Prospects

The monetary values for environmental impacts of emissions and resource use presented in the preceding chapters are still at an early stage of development. Like prices in the market they must be accepted by the concerned parties before having an impact in real life. Sustainable development is very much a societal concern, which means that governments are "concerned parties". For them to be able to accept the monetary values presented here, there is a need for scientific rigor and international harmonization. As far as possible I have tried to achieve this with the values presented here, but several factors contribute to uncertainties in the data obtained. One is the obvious problem of handling several thousand numbers varying from 1E–12 to 1E+12. A miscalculation can happen easily, and there is seldom something to compare the results with. The only way to decrease this problem is to be transparent so the errors, if occurring, can be detected in the future. A second factor is extracting relevant information from the literature. Each time I have made these updates, it becomes easier as more and more reviews are available. A third factor, and perhaps the most important one is what drives governments to be a "concerned party". Still today, CBA and LCA is required to be made for different public decisions, but the quality is seldom debated outside the walls of the experts performing the assessments. But as prices on goods from the environment go up, and they will, the willingness to pay for environmental qualities will increase. Hopefully, there will then be an increased consensus on which values to use for environmental goods and related anthropogenic impacts.

I believe a strategy towards achieving common data on impact values for emissions and resource use is to develop databases and software that decisionmaker and developers at all levels can use. Making monetary valuations of products and processes is a complicated task that only can be achieved if there is a cooperation between experts of several kinds. For that we need a certain degree of harmonization. A first step has been taken by the ISO 14008, and a second is coming with the standard ISO 14007. But more harmonization is needed, e.g., of the system boundaries in the analysis. Even if it seems impossible to find one solution for all, there can at least be a few main options.

A fourth factor is the completeness of the list of emissions and resources valued. A guiding principle is that those with high monetary values and low variability are to prefer. For instance, impacts from substances where there is an instant effect related to exposure are difficult to model without large uncertainties. But it is only when real cases are analyzed, that one can decide which information that was required or not. An example can illustrate the dilemma of including uncertain impact values: Assume that we have two products, one, A, with an emission of 1 kg of a substance and one, B, with an emission of 2 kg of the substance. The impact values of the substance vary with a factor of 10. A Monte Carlo simulation then shows that the probability that A has lower impact values than B is only 50.2%. Other simulations with lower uncertainties give somewhat clearer ranking, but the uncertainty need to be below 1.5 to allow a clear ranking of the alternatives (Figure 17.1). It may be intuitively felt that if there is a difference of a factor of F between two alternatives, the uncertainty, U, in terms of a standard deviation factor should not be larger than F and ideally U^2 should be less than F.

A particular problem is the valuation of impacts from PM2.5. PM2.5 has a large monetary impact value and its impacts vary with population density. As its residence time is around a week, the variation in population density among countries can be used as a measure of impact variability.

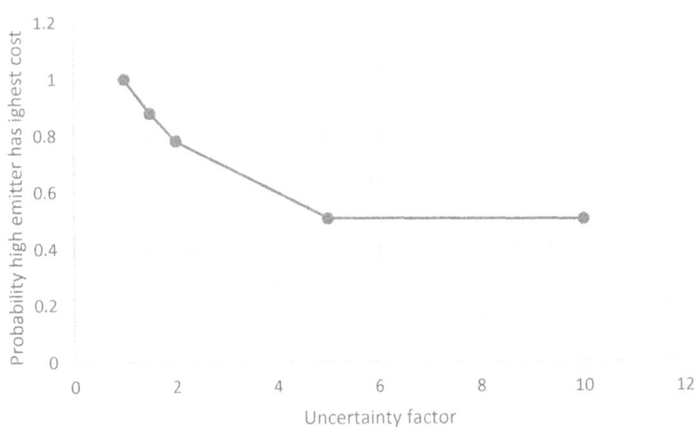

Figure 17.1. Result from Monte Carlo simulations of differences between two alternative products where one has twice the emission of the other.

The median population density in the world is 80 persons/km^2 according to the World Bank. 80 is also the average value in a log-normal distribution. The standard deviation is a factor of 4.4. 3.2 was assumed in the valuations in Chapter 7. To be a really useful measure for monetary valuation of environmental impacts, there is a need for regionalization of the values for PM2.5. As a first approximation, the ratio of population density in the country to the median (80) can be multiplied with the global average PM2.5 value to adjust it to a local condition. But also, the values for VOCs that are precursors to secondary aerosols need to be regionalized, in cases where VOC is important for the decision at hand.

There are a number of other emissions and resource extractions for which monetary impact values may be useful to have, and which are not valued in this book. The main reason is that uncertainty is high and average impact values moderate. In specific situation, they may still be valuable to have. Impacts from noise and emission of fertilizers to soil may be of much interest.

The values presented in this book have been intended to be used in development of technology and choice between product alternatives. But maybe the greatest value it has is for learning. When determining and using monetary values of environmental impacts it is inevitable that you reflect on why things are like they are and maybe you can add something new to the knowledge capital.

18

Summary

In Table 18.1, monetary values (costs) for decrease of environmental goods are summarized. In Table 18.2 monetary values of environmental impacts from emissions are summarized.

Table 18.1. Monetary values for environmental goods.

Environmental good	Unit	Monetary impact value ($)	
		Best estimate	Uncertainty
Production capacity of crops	kg	0.289	1.24
Production capacity for meat	kg	2.59	1.32
Production capacity for fish	kg	2.42	1.30
Production capacity for wood	m^3	72	1.2
Biodiversity	All, dimensionless	7.61E+10	1.5
Production capacity for drinking water	m^3	1.87	2.26
Coal resources	kg	0.284	1.3
Lignite resources	kg	0.084	1.2
Oil resources	kg	0.8	1.3
Natural gas resources	kg	0.5	1.3
Metal resources			
Aluminum	kg metal	0.175	1.3
Iron	kg metal	1.1	1.6
Silver	kg metal	115200	2
Arsenic	kg metal	3840	2

Table 18.1 contd. ...

... *Table 18.1 contd.*

Environmental good	Unit	Monetary impact value ($)	
		Best estimate	**Uncertainty**
Gold	kg metal	3200000	2
Bismuth	kg metal	44308	2
Cadmium	kg metal	58776	2
Cerium	kg metal	90	2
Cobolt	kg metal	226	2
Chromium	kg metal	43	2
Cesium	kg metal	1252	2
Copper	kg metal	144	2
Dysprosium	kg metal	1646	2
Erbium	kg metal	2504	2
Europium	kg metal	6545	2
Gallium	kg metal	339	2
Gadolinium	kg metal	1516	2
Germanium	kg metal	3600	2
Hafnium	kg metal	993	2
Mercury	kg metal	85970	2
Holmium	kg metal	7200	2
Indium	kg metal	115200	2
Iridium	kg metal	261818182	2
Lanthanum	kg metal	192	2
Lead	kg metal	339	2
Lithium	kg metal	288	2
Luthetium	kg metal	18000	2
Manganese	kg metal	5.6	2
Molybdenum	kg metal	3840	2
Niob	kg metal	480	2
Platinum	kg metal	9600000	2
Rhenium	kg metal	14400000	2
Rhodium	kg metal	320000000	2
Ruthenium	kg metal	192000000	2
Antimony	kg metal	28800	2
Scandium	kg metal	411	2
Samarium	kg metal	1280	2

Table 18.1 contd. ...

... Table 18.1 contd.

Environmental good	Unit	Monetary impact value ($)	
		Best estimate	Uncertainty
Tin	kg metal	698	2
Tantalum	kg metal	5760	2
Terbium	kg metal	9000	2
Tellurium	kg metal	5760000	2
Thorium	kg metal	2057	2
Titanium	kg metal	1	2
Thallium	kg metal	5760	2
Thulium	kg metal	17455	2
Tungsten	kg metal	5760	2
Uranium	kg metal	538	2
Vanadium	kg metal	54	2
Yttrium	kg metal	262	2
Ytterbium	kg metal	2618	2
Zinc	kg metal	46	2
Zirconium	kg metal	30	2
Alkali metal resources			
Lithium	kg metal	5	2
Sodium	kg metal	0	1
Potassium	kg metal	0	1
Rubidium	kg metal	0	1
Cesium	kg metal	780	2
Alkali earth metals			
Beryllium	kg metal	9600	5
Magnesium	kg metal	0	1
Calcium	kg metal	0	1
Strontium	kg metal	0.2	2
Barium	kg metal	11	2
Non-metal resources			
Hydrogen	kg	0.44	2
Boron	kg	10	2
Carbon	kg	0.284	1.3
Silicon	kg	0	1
Nitrogen	kg	0	1

Table 18.1 contd. ...

... Table 18.1 contd.

Environmental good	**Unit**	**Monetary impact value ($)**	
		Best estimate	**Uncertainty**
Phosphorous	kg	8.2	2
Oxygen	kg	0	1
Sulphur	kg	0.2	2
Selenium	kg	50	5
Halides			
Fluorine	kg	6	2
Chlorine	kg	0	1
Bromine	kg	0	1
Iodine	kg	30	2
Noble gases	kg	0	1
Helium	kg	0	1
Argon	kg	0	1
Neon	kg	0	1
Krypton	kg	0	1
Xenon	kg	0	1
Human health			
Working capacity	personhour	30	1.25
Years of lost life expectancy, YLL	personyear	107067	1.3
Undernutrition	personyear	6424	1.1
Diarrhea	personyear	11242	1.5
Malaria episodes	personyear	20450	1.1
Gravation of angina pectoris	personyear	6424	1.5
Cardiovascular disease	personyear	10707	2.4
Infarcts	personyear	8795	1.3
Asthma cases	personyear	4604	2
Chronic obstructive pulmonary disease, mild and moderate	personyear	18201	2
Cancer	personyear	21413	2
Skin cancer	personyear	5353	2
Low vision	personyear	18201	2
Poisoning	personyear	64240	1.2
Intellectual disability: mild	personyear	3319	4
Osteoporosis	personyear	137045	3
Renal dysfunction	personyear	68523	2

Table 18.2. Monetary environmental impact values for emissions.

Emission or resource	Unit	Monetary impact value ($)	Note
Carbon dioxide, CO_2 to air	kg	0.356	
Carbon monoxide, CO, to air	kg	1.33	
Nitrogen oxides, NO_x, to air	kg as NO_2	–32.8	
Nitrous oxide, N_2O, to air	kg	94.6	
Ammonia, NH_3, to air	kg	–53.2	
Sulphur oxides, SO_x, to air	kg as SO_2	–10.4	
PM2.5, to air	kg	268	
PAH in particles to air	kg	5.31	In addition to PM2.5
Halogenated organic compounds to air	kg	0.3 – 4340	Summarized in Table 8.1
Volatile organic compounds, VOC, to air	kg	0.41–43.8	Summarized in Table 9.2
Methane to air	kg	10.13	
C-14	TBq	1.20E+07	
H-3	TBq	1.70E+02	
I-129	TBq	5.65E+05	
Kr-85	TBq	2.83E+02	
Pb-210	TBq	1.41E+05	
Po-210	TBq	1.41E+05	
Ra-226	TBq	8.48E+04	
Rn-222	TBq	2.12E+03	
Th-230	TBq	4.24E+06	
U-234	TBq	1.13E+06	
U-238	TBq	9.89E+05	
Use of pesticides	kg	8.26 – 79600	Summarized in Table 11.2
BOD to freshwater	kg O_2	0.000352	
N_{tot} to freshwater	kg N	0.0024	
N_{tot} to seawater	kg N	0.006	
P_{tot} to freshwater	kg P	0.0455	
As to freshwater	kg	8030	
Cd to freshwater	kg	26200	
Hg to water or air	kg	435	
Littering on land	kg	5	
Littering to sea water	kg	4.25	

Table 18.3. Monetary valuations of impacts from land use.

Lan used activity	Unit	Monetary impact value ($)	Note
Residential & commercial developments in cities > 0.5 million inhabitants			
Housing and urban areas, arable land	m^2 year	10.52	
Housing and urban areas, forestland	m^2 year	10.39	
Housing and urban areas, impediment	m^2 year	10.35	
Commercial & industrial areas, arable land	m^2 year	10.51	
Commercial & industrial areas, forestland	m^2 year	10.38	
Commercial & industrial areas, impediment	m^2 year	10.34	
Tourism & recreational areas	m^2 year	0.13	
Residential & commercial developments in rural and cities < 0.5 million inhabitants			
Housing and urban areas, arable land	m^2 year	7.27	
Housing and urban areas, forestland	m^2 year	7.14	
Housing and urban areas, impediment	m^2 year	7.10	
Commercial & industrial areas, arable land	m^2 year	7.26	
Commercial & industrial areas, forestland	m^2 year	7.13	
Commercial & industrial areas, impediment	m^2 year	7.09	
Tourism & recreational areas	m^2 year	0.09	
Agriculture and Aquaculture			
Annual&perennial non-timber crops	m^2 year	0.00082	
Wood & pulp plantations	m^2 year	0.00152	
Livestock farming and ranching	m^2 year	0.00025	
Marine and freshwater aquaculture	kg produced	0.00477	
Energy production and mining			
Oil and gas drilling	kg produced	0.00004	
Mining and quarrying	m^2 year	0.624	
Renewable energy	m^2 year	0.00022	
Transportation and service corridors			
Roads and railroads	m^2 year	1.055	
Utility and service lines	m^2 year	0.043	
Biological resource use			
Logging and wood harvesting	m^2 year	0.00029	

Index

L

M

N

O

P

W

Y

For Product Safety Concerns and Information please contact our EU representative GPSR@taylorandfrancis.com
Taylor & Francis Verlag GmbH, Kaufingerstraße 24, 80331 München, Germany

www.ingramcontent.com/pod-product-compliance
Ingram Content Group UK Ltd.
Pitfield, Milton Keynes, MK11 3LW, UK
UKHW011000180926
475701UK00004B/481
* 9 7 8 1 0 3 2 0 8 6 2 4 8 *